AF443721

Chariots *for* Fire

CHARIOTS FOR FIRE

A PICTORIAL HISTORY OF THE NEW ZEALAND FIRE ENGINE

BRIAN DENTON

IPL BOOKS

First published in New Zealand in 1995 by
Reed Books,
a division of Reed Publishing (NZ) Ltd,
39 Rawene Road,
Birkenhead, Auckland 10.

First published outside New Zealand in 1995 by
IPL Books,
9 Cooper Street,
Smithfield, NSW 2164,
Australia.

ISBN 0 7900 0408 9 (New Zealand)

ISBN 0 908876 90 4 (Australia)

Production by IPL Books, Wellington, New Zealand.

Printed in Hong Kong through IPL Publishing Services.

Contents

Foreword

It gives me great pleasure to write the Foreword for a book that records the history of past and present fire appliances used in New Zealand.

Brian Denton is a true enthusiast who, through the compilation of this book, has dedicated hundreds of hours of his personal time to ensure our nation will always have a detailed photographic record of such fire appliances for future posterity. With his background I can think of no other person more qualified to put together a book that records the history of our past and present fire appliances.

Fire appliances generally hold a special and significant place in the early memories of most people. No doubt we can all recall how, as children, we would run to the gate or street corner to watch a big red fire appliance rush by with its siren wailing. Often we would jump on our bicycles, generally doubling a sibling or mate, and follow the appliance to get a closer look at the action as the fire-fighters attacked the flames, using 'the big red' to pump water.

As we flick through this book, many of our childhood memories will be revisited as we recognise that old fire appliance that we chased up the streets we then knew. Many of us never got past our childhood memories and finally joined the local Fire Brigade so we too could ride 'the big reds' and wave back to the many children standing on the street corners or following behind on their bicycles.

It is to this group of people that Brian dedicates his book — the past and present fire-fighters. The people who did, and still do, get out of bed in the middle of the night or leave their place of work and ride the appliances to assist their fellow human beings in their hour of need. The people who often torture their very own minds as they pick up the broken bodies of crash victims from the horrific motor vehicle accidents that sadly have now become very much a part of a fire-fighter's life. This book depicts the main piece of fire-fighting equipment used by all past and present fire-fighters and therefore stands as a fitting tribute to such a fine body of dedicated people.

Throughout the history of fire services in New Zealand, there has been considerable change in the administration of Fire Brigades. Partly because of such change there has been little or no comprehensive history programme instituted by the New Zealand Fire Service Commission, which came into existence in 1976. Brian and others have expressed considerable concern that many older and more picturesque appliances would be disposed of and lost for all time. It is for this reason that Brian has gone to all the trouble and personal cost of producing this pictorial record, consisting of some 350 photographs depicting fire appliances from the 1890s to the present time. In these photographs Brian has captured most, if not all, types of appliance used in New Zealand and has dedicated a section of the book to each of these categories. All past and present fire-fighters, whether volunteer or permanent, whether from urban, industrial or rural origins, will find appliances in this book that they will recognise and relate to and have fond memories of.

It is also a credit to Brian that any proceeds from this publication will go towards the establishment costs of a new museum that he and others intend establishing in the Okaihau area over the next few years. Surely this depicts the hallmark of a true enthusiast whereby any monetary reward that could have been personally gained is put back into providing a means of establishing a permanent record of our historical past. For this reason alone we should all support Brian and invest in this book.

On behalf of all fire-fighters of New Zealand, both past and present, I offer my congratulations and thanks to Brian Denton for his enthusiasm and initiative in producing this book which will now provide the nation with a comprehensive pictorial record of our past and present fire appliances.

Brian Stanley
President
United Fire Brigades of New Zealand

Introduction

The contents of this book represent many hundreds of hours of research and correspondence. I and my colleagues have, over the last few years, also visited personally a number of brigades in many areas to obtain photographs of elusive 'red flashing beasts'.

With the appliance replacement programme currently being undertaken by the New Zealand Fire Service, we were concerned that many of the earlier appliances would disappear and an important part of the automotive history of New Zealand would disappear with them.

Our objective was to create a pictorial history of as many of the appliances as possible that have been used in New Zealand, from the beginnings to the present day: to have some form of record for future generations of the amazing development of fire appliances over the years; to be able to compare the primitive technology of the early steam-powered vehicles with the latest turbo-driven appliances.

To many people a fire engine is just a fire engine. But to those who have been involved in fire-fighting in some way or other, they have played an important part in their lives. They come in all shapes and sizes and, contrary to popular belief, they are not always red.

We hope, as you browse through this book, that you get as much enjoyment and pleasure as we did in putting it together, and that this will go some of the way towards preserving the history of the New Zealand fire engine.

Brian R. Denton

SECTION ONE
From the beginning to 1940

Few people can resist the sight of a fire engine in full cry, with lights flashing and sirens wailing as it thunders its way to the scene. They are symbols of power, excitement and courage. They each represent the progressive ingenuity of mankind in the confrontation of one of the most persistent and feared adversaries, fire.

To the general public, a fire engine provides a sense of security and knowledge that we are all protected by a dedicated group of professionals or volunteers in each of our communities. To the younger generation, it embodies a group of heroes to be admired and represents that common childhood aspiration to be a fire-fighter in adult life.

But what is a fire engine?

Fire engines as we know them today first appeared around 1903, with Wanganui receiving the first self-propelled fire engine in Australasia. Before that, and indeed up until the beginning of the First World War, hand-drawn hose reels were in use by many brigades.

There were imported handpumps and later horse-drawn pumps. Some of these were in use about the early 1860s. Most of them required a lot of manpower — often ten to sixteen people were needed to operate the pump to get just a small amount of water onto a fire. The introduction of the horse-drawn steam pumpers took place in New Zealand in 1865 and the first sales were made to Wellington and Christchurch. Other brigades followed suit over the next few years. While these were a big improvement over the hand pumps, the main problem was the lack of water. They also required a skilful operator, able to get a good fire going and raise a head of steam. It could be seven to twenty minutes before the pump was able to be used at the scene.

From about 1910 the brigades began to import motor vehicles such as the Merryweather hose and ladder tenders.

The early appliances were of two types. The first and most common were the hose and ladder appliances. These were light and simple and were often adapted Model T Fords. They provided a reliable, fast method of transporting the fire-fighter and basic equipment such as ladders and hoses to the scenes of fires. The second were the pumps. These were heavier, slower appliances and more expensive. They were therefore fewer in number. In 1914, of the nine appliances held by the Auckland Fire Board, only one was a pump.

Most of the earlier appliances did not have the capacity to carry water. With improvements to the motor vehicle, the bigger and more powerful motors were able to propel bigger appliances.

The equipment continued to develop and brigades obtained better and faster appliances. Improvements also saw the introduction of specialist appliances such as self-propelled turntable ladders. An example is the 1912 Christchurch ladder, while some brigades purchased the heavier pumps which transported the wheeled escapes. The 1929 Leyland of Invercargill carried an 18-metre Simonis escape. As well as the bigger equipment, they also purchased smaller, lighter vehicles such as vans and even motorcycles to carry out some of their inspection duties.

The earlier appliances used the 'Braidwood' style of bodies whereby the fire-fighters perched precariously along the side of the appliance with their feet on the running boards. This gave them little or no protection and on a few occasions it was said that some unfortunates were thrown from the appliances. In the late 1930s, the new-style bodies started appearing, where the crew actually sat facing each other inside the rear section.

The first fully enclosed vehicle appeared in early 1943 when Hastings took delivery of a 1942 Ford fitted with a rear-mounted 30 L/S gear pump.

1846 TILLEY

Hand Pump
Horse-drawn pump
Estate model
Body mfct: Tilley (London)

This Tilley four- to six-man hand pump, owned by the Thames Volunteer Fire Brigade, dates from 1846. It was originally a hand-drawn vehicle but was later converted to be drawn by horses. It has cast iron wheels and no suspension, which no doubt provided a rough ride for those on board.

1860 MERRYWEATHER PUMP

Hand Pump

A 19th-century appliance, manufactured by Merryweather of England, consisting of a hand pump on steel wheels. It was pulled to where it was needed and usually required four strong men to operate it. After discovery in the small North Island town of Manaia it was restored by the members of the Fire Service Historical Society and is now on display in Christchurch.

MID-1800s SHAND MASON HAND PUMP

Built by Shand Mason, this manual pump was in use until 1921 in mid-Canterbury. A hand-drawn pump required up to twenty people, ten a side, to operate it. This was once used by the Fairlie Fire Brigade and is now on display at the Fire Service Historical Society in Christchurch.

1896 JOHNSON LADDER

Horse-drawn Ladder

Christened 'Vulcan', this ladder was built in Christchurch by P.M. Johnson based on the Shand Mason design for the Christchurch Fire Brigade. Firstly horse-drawn and later adapted to be towed by a motor-ised appliance, it was used until 1953 when it was sold to the Christchurch Parks Department. It was donated to the Fire Service Historical Society in 1970.

1910 HOSE REEL

This type of reel was widely used in the country and a few have been lovingly restored by a number of brigades. Note the teddy bear mascot on the draw bar. The photo, taken at Nelson and courtesy of the Alexander Turnbull Library, also well illustrates period uniforms worn.

1872 SHAND MASON STEAMER

Pictured is the second Christchurch steamer which was called 'The Deluge' and was stationed at the Chester Street Station (now Oxford Terrace). The unusual feature is the two hose reels towed behind the steamer. These appliances were often pulled by two or even three horses.

This steamer was originally purchased for the Kaiapoi Fire Brigade where it was used for a number of years. It was found in a quarry in Lyttelton Harbour and restoration started in 1969. It holds a current boiler ticket and is regularly put through its paces. It is on display at the Fire Service Historical Society.

1873 SHAND MASON

1887 SHAND MASON STEAMER

A further example of an early horse-drawn Shand Mason Steamer. This was once the pride and joy of the Masterton Fire Board. Known as the Jubilee pump, this steamer has since been stripped down and totally rebuilt by the Masterton Lions Club under the supervision of the Masterton District Council. Although it has a current steam certificate it will probably never be steamed up again as it is to be placed on permanent display in a special building adjacent to the local fire station.

1901 MERRYWEATHER STEAMER

Horse-drawn Steamer

Used by the Borough of St Albans and then by the Christchurch Fire Brigade, in 1921 this steamer was stationed at Smithfield Freezing Works near Timaru. Eventually it was passed on to the Fire Service Historical Society and restored. It was steamed up again in 1973. It is called 'Te Taniwha' which translated means 'The Water Dragon'.

1884 SHAND MASON STEAMER

Pump cap: 75 L/S
Horse-drawn Steamer

An appliance purchased by the Hastings Town Board in 1884 and christened 'The Deluge'. According to the local newspaper of the day, it could throw a jet of water higher than the tallest building there. The main problems with it were that it took up to 20 minutes to get up enough steam to operate the pump and there was often a lack of water due to the poor water supply. It is shown here decked out for a parade. Photograph courtesy of the Alexander Turnbull Library.

*Dennis Pump
Pump cap: 25 L/S*

Imported by the Christchurch Fire Board, this appliance arrived in New Zealand in January 1916. It was stationed at Christchurch Headquarters and later at Sydenham, Woolston and St Albans. In 1938 it was sold to the Islington Freezing Works who sold it in 1958 to a private owner. It is still basically in its original condition and runs well. Pictured at the Fire Service Historical Society in Christchurch.

1916 BUICK

*Body mfct: H Bates
of Napier*

Photographed at the Stratford Centennial Parade of Queen's Birthday weekend 1991, this belongs to the Southward Museum at Paraparaumu. It was once in service with the Hokitika Volunteer Fire Brigade, and also with the Granity Volunteer Fire Brigade.

FORD MODEL T

This appliance was purchased second-hand in 1920 by the Wanganui Fire Board. It was later sold to the Ohakune Fire Brigade. The photo is from the Telsa Collection of the Alexander Turnbull Library.

1917 STUDEBAKER
Reg. HI 9356

No Pump
Body mfct: H Bates
of Napier

Owned originally by the Dannevirke Fire Board, this Studebaker is now in the Southward Museum. Photographed at the Stratford Centennial Parade of 1991.

1926 FORD MODEL T

It is known only that this appliance operated in the Stratford District — it is not known precisely with which brigade. The photograph was taken by J.R. Wall of Stratford and is part of the J.R. Wall collection at the Alexander Turnbull Library.

1922 DENNIS
Reg. H66-388

Dennis Turbine Pump

Imported by the Wanganui Fire Brigade in 1922, this served there until the 1950s when it was sold to the Balclutha Fire Brigade who in turn sold it to a private collector. In 1972 it went up for auction and was purchased by the Fire Service Historical Society. It has since been restored.

1924 MODEL T
Reg. MO 7342

Photographed at the Stratford Centennial Parade in 1991, this Model T was in service for the Woodville Fire Brigade from 1927 until 1940 and was used to fight Woodville's biggest blaze on 6 August 1935, when fire razed the Club Hotel. The old appliance was discovered in Marton and has been privately restored.

1926 LEYLAND
Reg. DU 5117

Roturbo 10-193
Type 275 Pump

This appliance served all its life at Rangiora, 30 km north of Christchurch. It is now privately owned but is still under the control of the Rangiora Volunteer Fire Brigade. It is still in its original condition and is on display at the Fire Service Historical Society in Christchurch.

*Rees 8593 Type
275 Pump*

1922 LEYLAND

Purchased by the New Brighton Volunteer Fire Brigade in 1922 as their first new appliance, this was later sold to the Kaiapoi Volunteer Fire Brigade who had it in use until some time in the 1950s. It was bought by the Papanui Timber Company in 1964 for use at their sawmill at Franz Josef on the South Island's West Coast. It was provided on loan to the Fire Service Historical Society in 1974 and restored to its present condition.

Turntable Ladder (28 m)

1924 TILLINGS-STEVENS

Acquired in October 1924 for the Christchurch Fire Board at a total landed cost of £2844. In 1936 it was the sole exception to the board's decision to upgrade its fleet to pneumatic tyres. It was one of the machines used at the infamous Ballantynes department store fire in Christchurch in 1947. It was sold in 1963 to a demolition contractor and is currently privately owned but the Fire Service Historical Society has assisted in some restoration work.

1924 DODGE

This appliance was once the pride and joy of the Stratford Volunteer Fire Brigade and later used by the Toko Volunteer Brigade. It has been restored to its original condition and is now on display at the Stratford Pioneer Museum.

1928 DENNIS
Reg. ER 2164

Dennis Pump
Pump cap: 20 L/S

Purchased by the Wanganui Fire Brigade, this vehicle saw service there until it was sold in 1945 to a private owner. It has since been restored and featured in the Wanganui Fire Brigade's 125th Jubilee. It is photographed here in the Stratford Centennial Parade of 1991.

1928 LEYLAND
Reg. FE 114

Imported in 1929 for the Invercargill Fire Board, this was the first high-rise unit in New Zealand. It carries an 18.3-metre Simonis detachable water tower/escape. It weighs 7 tonnes and has a top speed of 68 km/h. It was purchased by the Ocean Beach Freezing Works at Bluff in 1957 and in 1977 was placed on permanent loan to the Fire Service Historical Society. It was in perfect condition with no need for restoration.

1930 MORRIS COMMERCIAL
Reg. EQ 1478

This 1930 appliance, built on a Morris Commercial truck chassis, was stationed at the Raetihi Fire Brigade and is now privately owned in Raetihi, still in its original condition. The photo was taken at the Stratford Centennial Parade in 1991.

1931 FORD AA

Hale Pump

Imported by the Lower Hutt Borough Council from the La France Corporation in 1931, this saw service with the Featherston Volunteer Fire Brigade until 1975, in later years as a ceremonial vehicle. It has been overhauled and restored to its original condition by the Fire Service Historical Society.

1931 DENNIS PE

Dennis Turbine Pump
Pump Escape

This appliance saw service with the Nelson Fire Brigade. During the early 1950s it was sent to Christchurch where the Bailey Escape was fitted and thus replaced the original 10.7-metre truss type ladder. Now privately owned, it is pictured on display at the Fire Service Historical Society at Christchurch.

A number of these appliances came to New Zealand for a variety of brigades. Purchased by the Te Awamutu Fire Board and commissioned on 28 March 1934, this one saw service for 48 years until 1981. The brigade then purchased it from the New Zealand Fire Service and, after a period of fundraising, completely restored it. To house it, a special building was constructed and opened in 1986.

1934 FORDSON V8

1935 DENNIS ACE
Reg. AS 7503

Purchased by the Christchurch Fire Board in 1935 and based at Sydenham station. In 1955 it was sold to the Kumara Volunteer Fire Brigade where it stayed until 1979. It was restored in 1982 by the Fire Service Historical Society in Christchurch.

1938 LEYLAND FK7
Reg. EW 3687

This is one of the only two appliances that were sent from the Leyland Works in England to New Zealand as a cab and chassis. Its body was built in Palmerston North where the appliance served (with the Palmerston North Fire Board) until 1955. It was then bought by the Glen Eden Borough Council for their brigade. In 1973 it was decommissioned, at which stage it had only travelled 10,090 miles (16,245 km). It is now at the Pioneer Village in Kaikohe in its original condition and still operational.

1937 DENNIS LIGHT SIX
Reg. DV 3434

Dennis Pump
Pump cap: 45 L/S
Pump
Body mfct: Stevens and Son (Christchurch)

The cab and chassis of this appliance were imported by the Christchurch Fire Board. The body was built in Christchurch to the local design and it served at most stations in Christchurch and finally at Lyttelton from 1967-1971. Following this it was restored by members of the Fire Service Historical Society and used for ceremonial duties.

1936 FORD V8
Reg. DV 2970

After starting life as a private car, this was converted in 1943 to a fire appliance by the Colonial Motor Company of Wellington. It served at a hospital near Christchurch until 1974 when it was donated to the Fire Service Historical Society in Christchurch.

1935 FORD V8
Reg. EZ 7190

This started service with the Dargaville Volunteer Fire Brigade and was used until 1956 when it was sold to the Silverdale District Fire Brigade. The latter rebuilt it and used it until 1970, from which time it served the Kumeu Volunteer Fire Brigade until 1972.

1939 LEYLAND
Reg. PP 5836

Pump

This was the second of the Leyland cab and chassis combinations to be sent to New Zealand from the Leyland Works. Purchased by the Masterton Volunteer Fire Brigade, it is unusual as the pump is mounted in the centre — up until that time most were rear-mounted pumps. It has since been restored and is in private ownership.

1940 FORD V8
Reg. DV 3439

Colmoco Pump
Pump cap: 33 L/S
Body mfct: Colmoco

Purchased by the Christchurch Fire Board from the Colonial Motor Company, this Ford served at the Christchurch Headquarters as well as at New Brighton. It was involved in an accident, finishing up on its side. After it had been re-paired, it was found to be very unstable, so a large piece of steel was put under the passenger's seat to correct the prob-lem. It is now in restored condition at the Fire Service Historical Society in Christchurch.

1932 MORRIS COMMERCIAL

Purchased by the Auckland Metropolitan Fire Board in 1932, this is said to have served in Avondale, Mt Albert and Mt Wellington. It was sold to the Whitianga Fire Brigade in 1955 and is now at the Historical Village in Tauranga. It was one of three of this type purchased by the Auckland Metropolitan Fire Board.

1924 CHEVROLET

Front Mounted Pump

Little information could be found on this appliance, but it is known to have once been used by the Leeston Fire Brigade. It is now in the ownership of a private collector.

1924 HUDSON
Reg. EQ 7122

Another appliance now in the Southward Museum in Paraparaumu. This Hudson was used by the Marton Fire Brigade. Photographed at the Stratford Centennial Parade in 1991.

1926 BUICK

The body of this appliance was built by the Gore Volunteer Fire Brigade on a Buick cab and chassis. One unusual feature is that it retained the left-hand drive. Little is known about its history, but it was given to the Fire Service Historical Society in 1979 for restoration and is now part of their collection.

Imported by the Dunedin Fire Board and served there for a number of years, during which time it was converted to run on pneumatic tyres. It was sold by Dunedin to the Springfield Volunteer Fire Brigade and is now owned by a private collector who has loaned it to the Fire Service Historical Society since 1983. The ladder gantries have been removed.

1927 LEYLAND
Reg. DV 5047

One of three of its type in the South Island, this served the Lyttelton Fire Brigade for all of its working life. The body was built by Steel Brothers of Christchurch. It was taken out of service in the 1960s and sold at an auction some time in the 1970s. Located by the Fire Service Historical Society with the pump and both axle shafts missing, as shown in the photograph, it now awaits restoration.

1938 DENNIS ACE

1934 FORD V8
Reg. EO 6455

Photographed here during the Stratford Centennial Parade, this appliance served the Inglewood Fire Brigade. Still in its original condition, it is now owned privately.

1935 INTERNATIONAL
Reg. CI 7286

Said to be the only one of its kind, this appliance was purchased by the Akitio County Council for the Pongaroa Volunteer Fire Brigade from the Greytown Volunteer Fire Brigade in 1964. It is now owned by a local farmer and used for parades and special occasions.

Nicknamed 'The Rattler', this was the first of the modern appliances to go into service with the Auckland Metropolitan Fire Board and became a legend for its speed from the station. Funds for its restoration were raised by some American sailors and it has pride of place in the MOTAT fleet, often used for funerals of brigade officers.

1935 FORD V8
Reg. EY 7797

Operated by the Wanganui Fire Brigade from December 1937, this was fitted with a 9.1-metre ladder and had two delivery outlets. A light six-powered appliance, it had a twin spark over-head camshaft petrol engine which is said to have been its best feature. Its handling left a little to be desired as although it is said to have 'gone like a rocket', it would not stop quickly. It was sold to the Waverley Brigade and then finally joined the MOTAT collection.

1936 DENNIS LANCE
Reg. EO 2861

1938 FORD V 8
Reg. ON HEAT

Colmoco Pump
Body mfct: Colmoco

Once operated by the Waitara Volunteer Fire Brigade, this is still owned in the area. It was photographed during the Stratford Jubilee Parade.

1938 FORD V8
Reg. EM 1803

Front mounted (make unknown)
Body: Standard Motor Bodies

Built for the Upper Hutt Borough Council by Standard Motor Bodies Ltd, this was transferred to the Upper Hutt Fire Board in 1943 and is now part of the collection at the Southward Museum. It has an unusual front-mounted pump.

Colmoco Pump

Photographed during the Kaikohe Golden Jubilee Parade, this Ford appliance was once stationed at Kawakawa Volunteer Fire Brigade and is now located at the Northland Regional Museum.

1939 FORD V8
Reg. DZ 268

Colmoco Pump
Body mfct: Colmoco

A good example of the ever-popular 1939 Ford appliance, a widely used design in the late 1930s. Many were still in service until the early 1970s. This example is seen here inside the station of the Dannevirke Volunteer Fire Brigade.

1939 FORD V8
Reg. FN 3518

1940–1960

In the late 1940s, the last of the open-type appliances, such as the Ford V8s, were starting service with many provincial city brigades. A number of brigades had this type as their main appliance and a few are still in use today in some industrial brigades and fire parties.

While these appliances gave little comfort to the fire-fighters, they were robust and were capable of carrying necessary equipment and crew to many incidents. Many were christened with various nicknames by their crews. These ranged from 'the Rattler' to 'the Admiral's Barge'.

During 1941 the Government of the day set up the Emergency Fire Service (EFS). This was an organisation expected to support the regular Fire Brigades. The members were given army-type uniforms and helmets and their main equipment consisted of trailer pumps. These could be towed by private cars, usually taxis.

Between 1941–1945 over 1600 of these trailer pumps were produced. Many of them were shipped overseas, but around 160 were kept in the country. Most of the heavier Dennis Trailer pumps were imported from England. These were put into service in the higher fire risk areas. These trailer pumps became widely used by many fire brigades after the EFS was disbanded. As the prices of the pumps were relatively low, they became the mainstay of the equipment used by a large number of the smaller volunteer brigades. They were usually towed behind a car or truck belonging to one of the volunteers.

At the close of the Second World War a number of surplus military vehicles became available, including second-hand crash fire trucks, four-wheel-drive vehicles, plus an abundance of military vehicles which could be easily transformed into fire appliances. A number of ex-Army Quads and Fargo trucks were soon appearing as combination units in many areas. Some were converted to forestry units and were still operational until the late 1980s.

It was about the same time that the first of the fully enclosed type appliances were starting to appear. The streamlined type — the Fargo and Ford V8 — with their fully enclosed body and lockers not only gave better storage for the equipment, but also more protection for the fire-fighters from the weather.

During the 1950s, the Dennis F12 and the smaller F8 enclosed appliances, with Rolls Royce motors and 'crash' gear boxes, began to arrive from England. Most of these went into service in the larger city brigades. The larger F12 was usually fitted with the wheeled escape ladders, and were a common sight in the larger urban areas. The smaller F8s were used widely in the four main centres as well as a number of provincial centres. The F8 became a popular appliance as its smaller wheel base suited the tight cornering required in many narrow city streets. Many a modern-day fire-fighter has served his 'apprenticeship' driving these.

Some of these Dennis F8s were later converted to specialist type appliances such as rescue tenders or foam tenders.

1941 FORD TRAILER PUMP

Colmoco Pump
Pump cap: 30 L/S
Trailer Pump
Body mfct:
Colonial Motor Co.

More than 1700 of these pumps were built during the Second World War by Colmoco in New Zealand and supplied to the Ministry of Defence and the EFS. Powered by a Ford V8 motor and coupled directly to the pump, these units became very popular with many volunteer fire brigades after the war.

1942 DENNIS TRAILER PUMP

Dennis Pump
Pump cap: 38 L/S

Imported from Dennis in England and supplied to the National Service Department which formed the EFS. After the war, about 1947, a number were handed over to local brigades. These were often towed behind appliances and some were still in use as late as the mid-1980s. The one shown was formerly at Christchurch and is now at the Fire Service Historical Society.

1954 GWYNNE TRAILER PUMP

A modern version of the popular trailer pump. Most were powered by a four-cylinder side-valve petrol engine. The actual pump unit was able to be removed from the trailer section to be located closer to the water source. The suction hoses were housed in the shaped wings in the trailer body. A special feature fitted to most was the folding handles on the draw-bar. This enabled the whole trailer to be moved manually as required.

1941 FORD PUMP

A skid-mounted version of the Colmoco pump, these were designed along the same lines as the trailer pump but for fitting in a permanent location. They were used on a number of the Fargo appliances as well as being located in certain high-risk buildings. They were powered by their own Ford V8 motor in direct drive to the pump. (Refer to page 38.)

4x4 Pump
Body mfct: Local

An ex-Army Chevrolet Quad converted to an appliance for the Rongotea Volunteer Fire Brigade in 1956, the year this brigade was formed.

1942 CHEVROLET QUAD

Simonis Pump
4x4 Pump
Body mfct:
Otahuhu Railway
Workshops

Adapted from an ex-Army ambulance and purchased by the New Zealand Railways for the Otahuhu Workshops in Auckland, the NZR coach-builders rebuilt it (as shown) as a first-aid tender. It was fitted with a Simonis pump powered by the PTO. It saw service at Otahuhu until it was transferred to Addington Workshops in 1982. In 1990 it was decommissioned and donated to the Fire Service Historical Society complete with its own Fire Station.

1942 CHEVROLET QUAD

1942 INTERNATIONAL 4 x 4
Reg. EZ 7178

Colmoco Pump
Pump cap: 38 L/S
Tank cap: 1820 L
4 x 4 Pump
Body mfct: Modified
by STD Motor Bodies

Originally this was built with an enclosed cab and a John Bean 'Royal' HP Pump. After use by the US Armed Forces in the Pacific, it was brought to New Zealand and stationed at Camp Bunn in Mount Wellington. It was given to the Auckland Metropolitan Fire Board in 1947, then in 1949 to the Waitemata City Council. Standard Motor Bodies in Wellington made modifications and removed the John Bean pump. It served for 27 years in the Waitakere Ranges before purchase and restoration by the Titirangi Volunteer Fire Brigade.

1942 FARGO
Reg. EU 3537

Once used by the Wellington Fire Brigade, this still carries the Wellington number 19 on the front. It is now in private ownership and has been restored. Photographed at the Stratford Centennial Parade.

Originally owned by the Internal Affairs and allocated to Gisborne, in 1967 this was sold to the Tokomaru Bay Volunteer Fire Brigade after having run 2634 miles (4240 km). It remained in service around the East Coast until it was decommissioned in 1979 when it was offered to the Fire Service Historical Society. It is still in its original condition. (Refer to page 35 for pump details).

1942 FARGO COMBINATION UNIT
Reg. EM 1812

A typical example of an appliance built for the EFS and then purchased by a number of volunteer brigades at the end of the Second World War.

1943 FARGO
Reg. EM 1866

1943 FARGO HOSELAYER
Reg. DV 3452

Colmoco Pump
Pump cap: 33 L/S

Originally a wartime 'combination unit', this appliance had hose lockers mounted behind the deck pump. In 1951 it was converted to carry a wheeled escape. This was removed in 1969 and the rear lockers were altered to form a hose layer. It also towed a Dennis Trailer Pump. After serving in the Christchurch area it was decommissioned in 1976 and is now at the Fire Service Historical Society.

1943 FARGO

Colmoco Pump
Body mfct: Colmoco Pump

The remains of this appliance were found in the Waitakere Ranges in West Auckland. It was a National Service Fargo purchased by the Auckland Fire Board in 1947, rebuilt by Eaddy and Taylor in 1954 to its streamlined shape and then sent to Mt Wellington in 1955. It was sent to Henderson to replace one of their appliances and also served at Te Atatu. Now resting quietly, it awaits restoration.

1942 FORD V8
Reg. EM 8099

Said to be the first fully enclosed fire appliance to be put into service in New Zealand, this had a rear-mounted pump. It was based at Hastings and remained in service for 30 years.

Colmoco Pump
Pump cap: 33 L/S
Body mfct: Colmoco

1946 FORD V8
Reg. ER 1769

This Ford began service with the Palmerston North Fire Board. In addition to the Standard Colmoco rear pump it also had a John Bean 'Royal' HP pump. In 1958 it was sold to the Feilding Fire Board where it served until 1980. It was restored by the Fire Service Historical Society in 1989.

1942 FORD V8
Reg. ET 9529

Issued to the New Zealand Army as NZ15622, this served at Trentham Military Camp until 1949 when it was transferred to the Upper Hutt Fire Board. In 1976 it was sold to the Waipaiata Youth Correction Centre and was stationed there until 1980, when it was transferred to the Paparua Prison in Christchurch. In 1984 it was handed over to the Fire Service Historical Society and restored.

1955 BEDFORD
Reg. EM 8101

This appliance, the first to be built to the New Zealand Fire Council specifications, started its service at Hastings and was also based at Toko in the Stratford area, before being withdrawn. It was fitted with a hose reel and a feather-weight pump.

1958 BEDFORD A2

Reg. DY 3002

Pump
Tank cap: 450 L

Only 25 of this style of appliance are understood to have seen service and little information is available on them, although it is known they were fitted with gear pumps and hose reels. They represented an inexpensive engine for small country towns. This example was once stationed at Governor's Bay Volunteer Brigade.

1957 BEDFORD A3

Reg. DT 4646

Renown Pump
Pump cap: 40 L/S
Tank cap: 1350 L
Medium Pump

This unusual long wheelbase Bedford was in service with the Ikamatua Volunteer Fire Brigade, an auxiliary brigade of Reefton on the West Coast of the South Island. Photographed awaiting disposal in the Christchurch workshops, it has since been replaced by a 1971 Ford.

1948 FORD V8
Reg. AS 2463

Colmoco Pump
Pump cap: 33 L/S

Not much is known about this appliance except that the body was built by the coach-builders at the Addington Railway workshops in Christchurch in the early 1950s. It was stationed at Otira until 1976 when it went to the Shanty Town museum near Greymouth. When found by the Fire Service Historical Society, it was said to have taken 18 months to 'dry out' as it had been abandoned outside. It was restored and used in the Otira reunion in 1989.

1955 DENNIS F8
Reg. DS 4900

Dennis Pump

This is said to have served in Timaru and Wakefield and been sold in the mid-1980s. During the early 1970s it was altered with the First Aid hose reels being removed and one being placed above the pump. The suctions were relocated to the roof area above the top lockers.

1957 BEDFORD 'GREEN GODDESS'
Reg. HR 318

Imported into New Zealand in 1975 and stationed with the Omakau Volunteer Fire Brigade, this Bedford was repainted red on arrival and, after NZFS nationalisation, orange. It was restored in 1984 by the Fire Service Historical Society and repainted the original green. Before coming to New Zealand it was owned by the Pembrokeshire Fire Brigade AFS in the UK and stationed at Tenby (UK Reg. NYV 566).

1954 BEDFORD VASS
Reg. HT 5612

Another example of the imported 'Green Goddess', this one saw service at Riversdale, painted in the NZFS colours, and was still operational in early 1992.

BEDFORD O SERIES
Reg. AE 3334

Hose Tender

This Bedford, thought to be about a 1946 model, was converted to a hose tender and saw use at the Conical Hill Fire Brigade. Little information is available about the brigade, but it is understood to have been known originally as the Conical Hill Sawmill Fire Brigade.

1949 BEDFORD K
Reg. FE 8251

Hose Tender

This early model Bedford was photographed at the Cambridge Volunteer Fire Brigade. It was used to tow the Ford V8 trailer pump and carried a range of hoses and equipment. Following its disposal by the brigade, the local borough council acquired it for use as a truck.

Deployed by Te Karaka
Volunteer Fire Brigade in
the Gisborne District, this
appliance, used in con-
junction with their old
1935 Ford V8, was built
by the brigade members
and the local garage of
Graham Motors. The
tank is an old under-
ground petrol tank fitted
on to a Bedford chassis.

1953 BEDFORD M
Reg. BX 1528

Hose Reel Tender

Another version of a
Bedford hose reel tender
as used by the Eketahuna
Volunteer Fire Brigade,
this one was also used to
tow a Ford V8 trailer
pump. It has since been
restored in private
ownership.

1952 BEDFORD MD
Reg. LN 6025

1956 BEDFORD A3
Reg. EY 1091

A number of these appliances were imported from England fully built on a Bedford A3 chassis. This one was in service with the Ngahere Volunteer Fire Brigade in the Greymouth region. It has since been disposed of.

1957 BEDFORD A3
Reg. OM 512

Dennis Pump
Pump cap: 30 L/S
Tank cap: 450 L
Body mcft: Prestige

Photographed at the Stratford Centennial Parade, this Bedford was once stationed at Woodville in the Palmerston North region. It is now privately owned and fully restored.

This was the first appliance used by the Ashley Clinton Auxiliary Volunteer Fire Brigade in the Hawke's Bay area. It now has an International D200 appliance.

1954 BEDFORD
Reg. EN 2428

Pump

This started service at the Westport Volunteer Fire Brigade and also served other West Coast brigades. At the time of this photograph at the Christchurch workshops it was with the Whataroa Volunteer Fire Brigade. It was sold in 1989, fully operational.

1958 BEDFORD S
Reg. ES 9521

1961 BEDFORD J3
Reg. HI 4982

*Body mfct: Local
Rural Fire Appliance*

This appliance was built for the Rongotea Volunteer Fire Brigade by the local county council to replace their early model ex-Army Quad appliance.

1958 AEC MERRYWEATHER

*Merryweather Pump
Tank cap: 675 L
Body mfct: Imported
built up body*

One of two appliances built for the North Shore Fire Board and imported from the United Kingdom in 1958, these were the first diesel-powered pumps to go into service in the Auckland area. The two were Reg. EX 9217, sold to Fiji in 1983, and Reg. EX 9218, sold to Palmerston North.

1958 KARRIER GAMECOCK
Reg. DY 3004

Photographed on station at Renwick, this is one of many Karrier appliances that represented the mainstay of volunteer fire brigades in the late 1950s and early 1960s. Some were still operational as recently as 1991.

1958 KARRIER GAMECOCK
Reg. DV 3437

This was said to have served in Lyttelton and Rolleston, then at Havelock. It has polished metal sides on the cab and rear lockers. What appears to be an enclosed tray for the suction hoses built on the top of the locker section gives the rear an unusual appearance. It was sold in 1988.

1959 KARRIER CARMICHAEL
Reg. DS 2354

Gwynne Pump
Pump cap: 38 L/S
Tank cap: 900 L
Medium Pump

Another Karrier appliance, stationed at Waimate in the South Island, this particular one has an unusual body design. The entry to the cab is through a sliding door on the side instead of the usual hinged driver's door and folding crew cab door. It also has a lift-up locker door in place of the usual roller doors. Compare this with the other Karriers on page 50.

1948 MORRIS
Reg. FE 5485

Built from a Morris truck cab and chassis by the Whangamata Volunteer Fire Brigade, this appliance entered service in 1950 and served until replaced by a new Ford appliance in 1972. It is affectionately called 'Old Heartache'.

1957 COMMER CARMICHAEL
Reg. EH 5143

After beginning life with the Otorohanga Fire Brigade, in the mid-1980s this vehicle was sent to Colville Brigade on the Thames Peninsula. In 1991 the Thames District Council offered it back to the Otorohanga District Council Rural Fire Authority who accepted. It has covered only 7500 miles (12,100 km) and is still in its original condition.

1957 COMMER

This appliance was brought to New Zealand in 1957 to equip the new Fire Service training school at Island Bay in Wellington. The school was set up as part of the recommendations after the fatal Ballantynes fire of 1947, with another recommendation being to introduce to the larger brigades a wheeled escape to enable upper floors in multi-storied buildings to be reached. A wheeled escape was in operation in Auckland until 1971. MOTAT is where this appliance is to be found today.

1951 DENNIS F12

Reg. DV 3443

Dennis Pump

Imported by the Christchurch Fire Board in 1952, this first example of an F12 in New Zealand was fitted with a 15-metre Bayley Escape which had previously been imported in 1949. It served as a pump/escape, pump and training pump until 1981. It is now with the Fire Service Historical Society.

1953 DENNIS F12

Reg. EU 3520

Dennis Pump
Pump cap: 67 L/S

The Wellington Fire Board imported this appliance in 1953 and it spent most of its time in the Wellington area. It was transferred to Christchurch in 1977 for personnel training, remaining there until 1979. It was then given to the Fire Service Historical Society and fully restored in 1985.

Dennis Pump

One of several Dennis appliances imported by Auckland's Metropolitan Fire Brigade, this began service in 1955 at Ellerslie. In 1960 it was transferred to Mt Roskill, then to Onehunga in 1964. In 1965 it was converted to a Foam Tender and based at Pitt Street, then in 1967 at Parnell and Otahuhu. In 1979 it was downgraded to relief status. In 1982 it was given to MOTAT and has since been restored in its original status as a pump.

1955 DENNIS F8
Reg. EZ 1863

Dennis Pump

Despite the nominal model year, this appliance arrived from Britain on 14 September 1954. It was first stationed at Christchurch Headquarters, St Albans and New Brighton before being transferred in 1973 to Brooklands on the establishment of that station. In 1983 it was transferred to the Fire Service Historical Society and is still in excellent order.

1955 DENNIS F8
Reg. DV 3445

1956 DENNIS F8
Reg. 2SL04U

Dennis Pump

This appliance was originally stationed at Dunedin as part of a large re-equipment programme carried out in the 1950s by the Dunedin Fire Board during which a number of Dennis appliances were purchased. It later served at Dipton Volunteer Fire Brigade and is now at the Yaldhurst Fire Museum in Christchurch.

1957 DENNIS F8
Reg. DV 3448

Dennis Pump

One of the four F8s imported into Christchurch in the 1950s, this saw service in Christchurch, New Brighton, Lyttelton and Kirwee. It was sold in 1988, fully operational.

This appliance was a wartime Ford SP towing appliance. It was obtained by the Stratford Volunteer Fire Brigade who used it to tow a Ford V8 trailer pump. A pump was later fitted to the appliance and it was used until it was replaced by a 1960 Karrier. It has been restored and is now owned by the Egmont Rod and Custom Club at Stratford. It is photographed here at the Stratford Centennial Parade.

1944 FORD V8
Reg. EO 9883

Colmoco Pump
Body mfct: Colmoco

This Ford, one of the last of the open style appliances built in 1948, was in service at Ruakaka, just south of Whangarei, between 1970 and 1979. A section of special lockers have been built on the rear section to store the equipment carried. Extra lengths of coiled hose have been placed in a special rack on the rear step of the vehicle. Ruakaka Brigade is the first back-up brigade for the New Zealand Oil Refinery at Marsden Point.

1948 FORD V8
Reg. EK 2347

1949 FORD V8
Reg. AL 1690

The last Ford V8 purchased by the Christchurch Fire Board, this arrived in 1949 and was stationed at Christchurch Headquarters. In 1956 it was sold to the Kaiapoi Fire Board and then again in 1971 to the Dunsandel Volunteer Fire Brigade. The latter used it as a rescue unit and a pump. After being handed over to the Fire Service Historical Society in 1985, it was stripped, rebuilt and repainted.

1952 FORD V8
Reg. ER 4587

A type of appliance popular with small communities, this example was purchased by the Waikanae Volunteer Fire Brigade and served there until 1981. It carries 300 litres of water fed through the first-aid hose reel. Still in its original condition, it is now at the Fire Service Historical Society.

1953 FORD V8
Reg. EW 6043

An earlier Ford V8 built by the Colonial Motor Company in Wellington, this appliance served with the Howick Volunteer Fire Brigade until decommissioning when it was given to MOTAT. The Howick Volunteer Fire Brigade have been involved with restoring this appliance, photographed here on display during the 'open day' organised in Auckland. It has been re-registered RP 9593.

1955 FORD V8
Reg. DV 8379

This appliance served with the Kaikohe Volunteer Fire Brigade from 1955 to 1974. It was then sold to the Russell Volunteer Fire Brigade and is still used occasionally. It has been the main pump used at some of the larger fires around the Mid-North such as those at the Kaikohe War Memorial Hall, the Rangiahua Hotel and the Junction Hotel at Kawakawa.

1955 FORD V8
Reg. DA 4851

Front Mounted Pump
Pump cap: 28 L/S
Tank cap: 60 L
Light Pump
Body mfct: Canadian Ford

This short-wheelbase Ford was in service with the Papatoetoe Volunteer Fire Brigade from 1956 to 1975 — until 1968 as a pump, then it was converted to a hose layer. It carried 18 lengths of 90-mm flaked hose. Due to problems in obtaining spare parts it was withdrawn, restored and donated to MOTAT.

1959 FORD F600
Reg. DY 1342

Coventry Climax Pump
Pump cap: 40 L/S
Tank cap: 900 L
Body mfct: Colmoco

Built by the Colonial Motor Company Ltd in Wellington, this was designed by the members of the Picton Volunteer Fire Brigade and purchased by the Picton Fire Authority in 1959. It served at Murchison from 1980 to 1987 and was then restored to its original condition by the Picton Volunteer Fire Brigade. It is now used only for funerals or historic occasions.

1953 LANDROVER SERIES 1
Reg. DT 813

Pegson Pump
Pump cap: 15 L/S

This appliance was supplied to the Ashburton County Council in 1953 and was stationed at the small country town of Methven where it remained in service until 1984. It was transferred to the Fire Service Historical Society in its present condition with only 7000 miles (11,300 km) on the clock.

1956 LANDROVER SERIES I
Reg. AD 2934

KSB Pump

Supplied new to the Clyde Brigade in 1956, this appliance received some modifications during its service there. The first-aid hose reel was moved from behind the cab and fitted to the front bumper; the headlights were shifted from the grille to the mudguards; the pump was altered from a 75-mm to a 100-mm type; the 100-mm suction hoses were put into a tray fitted on the roof next to the ladders (the 75-mm suction hose brackets are still on the front mudguards). It also carried a portable pump on the rear deck between the side lockers. The Waipori Falls Brigade acquired it with only 2400 miles (3900 km) on the clock.

1958 LANDROVER SERIES I
Reg. DX 5368

Coventry Climax Pump
Light Pump

This is a good example of the ever-popular Landrover fire appliance used by several brigades for many years. Some of them are still in active service in rural areas. Able to carry equipment and manpower to many almost inaccessible areas, the short wheelbase version quickly became important to these brigades. This appliance was photographed in service with the Richmond Fire Brigade.

1958 LANDROVER SERIES 1
Reg. FLICK

Purchased by the Hokianga County Council in 1958 this was later sold to the Mangonui County Council and supplied to the Ahipara Fire Brigade. It carried a Wajax portable pump on the front and a Tohatsu larger portable pump between the rear lockers. Since acquisition by the Ahipara Brigade it has been repainted and is used in conjunction with the Far North Surf Rescue on Ninety Mile Beach.

1959 LANDROVER SERIES II
Reg. EN 3517

The long wheelbase version of the Landrover was just as popular an appliance as the shorter version. Many were built in England by Landrover and Carmichael and some saw service in New Zealand. This one was stationed at Waipukurau in Hawke's Bay. It carries a first aid reel, standard standpipe and the suction hose coiled around the bonnet. It has now been sold to the Hawke's Bay District Council.

1960 LANDROVER SERIES II
Reg. DW 8615

A long-wheelbase Landrover that has been stationed at New Brighton and Brooklands, this has a small water tank, pump and first-aid reel. The suction is also carried around the front bonnet. Most of these appliances have since been sold and a number have been purchased by local fire parties.

SECTION THREE
1960–1976

During this time the Karrier Gamecock, a cheaper machine than those that had appeared hitherto, became available. These were fitted with Karrier Truck or Humber Super Snipe motors. Their aluminium and wooden bodies suited New Zealand conditions and they were sought after by many of the smaller volunteer brigades. These were soon followed by the Commer Carmichaels which were the mainstay of the fire-fighting force until the arrival of the Fords and Bedfords in the late 1960s.

As well as the fully assembled Commers, Bedford, Ford and Commer cab and chassis sets were being imported, and companies such as New Zealand Motor Bodies, Colonial Motor Company and Wormald designed and built bodies for them. These appliances were forerunners of the 'new breed' of vehicles specially built for New Zealand brigades.

Other vehicles that proved to be very popular were the four-wheel-drive Landrovers. Both the series one and series two models were used by a number of rural brigades. Some were specially imported in the built-up stage complete with water tanks and pumps, while others were converted to suit local needs.

A number of Bedford VASS four-wheel-drive appliances were also imported from England. Many of these appliances had been in storage during the 1950s, awaiting the nuclear attacks that never eventuated. These rugged machines with their powerful 8 L/S pumps were sought after by a number of rural brigades. They were often called 'Green Goddesses' owing to their dark green military colours on arrival.

In the 1970s, a bigger variety of Fords became available and a number of the bigger city brigades were starting to obtain International appliances. The square shape of the body on the early models caused many to be nicknamed 'the butter-box'. The later models had bodies built by Mills Tui and Wormald on an imported cab and chassis, and this led to a variety of designs in storage and lockers.

1962 AEC
Reg. EX 9219

Built in New Zealand by Wormald, based on the design of the Merryweather as previously acquired by the North Shore Fire Board, this appliance was based at East Coast Bays for many years until being transferred to Mercer station. Following its involvement in a motor vehicle accident, it was written off.

1960 KARRIER GAMECOCK
Reg. DU 4280

A popular appliance with many of the volunteer brigades, this example was in service at Waipara when this photograph was taken and was still in service in 1990 after 30 years of operation.

1961 COMMER GAMECOCK
Reg. ET 8322

Gwynne Pump
Pump cap: 38 L/S
Tank cap: 900 L
Medium Pump

Originally a North Island machine, now stationed at Dunsandel in the South Island, this appliance is said to have been repowered with a 308-V8 motor. It was still in service in 1990.

1963 KARRIER GAMECOCK
Reg. DX 4229

Gwynne Pump
Pump cap: 38 L/S
Tank cap: 900 L
Medium Pump

Another Karrier stationed at Upper Moutere, still operational in 1991.

Gwynne Pump
Pump cap: 38 L/S
Tank cap: 900 L
Medium Pump
Body mfct: Roy
Chapman, Nelson

This appliance was originally built to carry a wheeled escape, but this has since been removed. Shown here at Rai Valley, it has also been based at Nelson and Stoke. This is the only one of its kind known to have been built especially with the longer wheelbase and this rear locker design. It has been withdrawn from service and sold to a local company for promotional use around Nelson.

1965 COMMER VAKS
Reg. CT 3888

Gwynne Pump
Pump cap: 38 L/S
Tank cap: 900 L
Medium Pump

This appliance was stationed at Greymouth and also at Ngahere. It became a relief pump there before being acquired by the Whangapoua Rural Fire Party on the Coromandel Peninsula.

1966 COMMER VAKS
Reg. DX 1200

1966 COMMER
Reg. DN 2501

Gwynne Pump
Pump cap: 38 L/S
Tank cap: 900 L
Medium Pump

Photographed while stationed at Weston, this appliance served a number of brigades in Region 6 in the lower part of the South Island. This particular appliance has a polished cab and rear locker section. It was sold in 1988.

1968 COMMER CV
Reg. DX 9413

Dennis No. 2 Pump
Pump cap: 38 L/S
Tank cap: 900 L
Medium Pump
Body Mfct: Roy
Chapman, Nelson

This Commer served a number of brigades in Region 3B. It has the typical polished rear cab and lockers seen on these types of appliance. It is photographed here as a relief pump in Wanganui in 1993.

1969 ERF 84PF
Reg. FH 2397

Godiva Pump
Pump cap: 38 L/S
Tank cap: 900 L
Medium Pump
Body mfct: Wormalds

The first ERF to go into service in New Zealand, this was based at Otorohanga in the North Island. It has since been replaced by a 1991 Iveco.

1971 ERF 84PF
Reg. DJ 8817

Godiva Pump
Pump cap: 75 L/S
Tank cap: 1350 L
Heavy Pump
Body mfct: Wormalds

Another version of the ERF appliance that began service in Christchurch; when photographed it was stationed at Ashburton. They were built with fibreglass cabs. Initially they were used by larger brigades; upon being superseded they were downgraded to volunteer pumps. This one has a closed crew cab and hinged lockers.

1971 EFT 84PF

Reg. FS 5279

Godiva Pump
Pump cap: 75 L/S
Tank cap: 1350 L
Heavy Pump
Body mfct: Wormalds

Received in 1985 by the Lincoln Volunteer Fire Brigade upon its upgrade to a two-pump station, this was previously based in Christchurch where it started service with the Christchurch Metropolitan Fire Board. It has the closed crew cab and polished metal locker section.

1972 ERF 84PF

Reg. FZ 1798

Coventry Climax Pump
Pump cap: 75 L/S
Tank cap: 1350 L
Heavy Pump
Body mfct: Wormalds

One of a group of ERF appliances used throughout the country, powered by a Perkins V8 diesel motor and with an open crew cab. This appliance is stationed at Ngatea on the Hauraki Plains.

Dennis Pump
Body mfct: Roy Chapman
and Company,
Nelson

Bought by the Dunedin
Metropolitan Fire Board
after having a body built
onto it in Nelson, this
served in Dunedin and in
a number of Southland
Brigades before being
given to the Fire Service
Historical Society in
1989.

1964 DENNIS F28
Reg. DK 2949

Dennis No. 3 Pump
Pump cap: 55 L/S
Tank cap: 680 L
Heavy Pump
Body mfct: Dennis, UK

Another Dennis appliance
bought new by the
Dunedin Metropolitan
Fire Board, this carried a
wheeled escape for many
years. It was moved to the
Darfield Volunteer Fire
Brigade in 1987 where it
was stationed until 1990
when it was replaced by a
D Series Ford. It was sold
to the Fire Service Historical Society who are
presently endeavouring to
obtain a wheeled escape
to fit onto the machine
again.

1964 DENNIS F106
Reg. DK 2951

1968 DENNIS
Reg. ED 8768

Dennis Pump
Pump cap: 38 L/S
Tank cap: 450 L
Body mfct: Dennis, UK

This appliance was supplied new to the Wellington area and at one stage was stationed at Plimmerton before being moved to the Wanganui area. The photograph was taken while the appliance was a relief pump at the Wanganui station.

1975 DENNIS F49
Reg. HW 4389

Dennis No. 3 Pump
Pump cap: 76 L/S
Tank cap: 450 L
Body mfct: Dennis, UK

Photographed in the rear of Wellington Central station in 1987 when it was a relief pump, this Dennis was purchased by a private Christchurch collector in 1989.

An unusually shaped
Bedford, this was
photographed in service
at Ward in the Blenheim
area.

1961 BEDFORD CSL23
Reg. DX 3018

This Bedford was built
on a longer than
normal chassis as a trial
unit, the only one
known to have been
so adapted. It was
variously stationed at
Henderson and
Ngaruawahia and as a
reserve pump at Taupo.
Seen here at Turangi.

1965 BEDFORD KDS
Reg. HO 8234

1966 BEDFORD TK

Reg. CU 8848

Coventry Climax Pump
Pump cap: 34 L/S
Tank cap: 900 L
Medium Pump
Body mfct: Wormalds

A large number of these appliances were scattered around many volunteer brigades and they are still used by several. This one was stationed at Mamaku in the Rotorua area.

1966 BEDFORD TK

FWP Pump
Pump cap: 20 L/S
Tank cap: 1800 L
Medium Pump

This was one of the two Bedfords stationed at Waikari in North Canterbury. Both have since been replaced by new Hino appliances.

Hale PTO Pump
Tank cap: 1800 L
Light Pump
Body mfct: NZ Forest Service

This appliance was originally built and owned by the NZ Forest Service, and stationed at Rai Valley. The Hale pump was replaced by a Wajax pump due to the shortage of parts for the Hale. Photographed at Tapawera, it was still in service in 1991.

1966 BEDFORD J2
Reg. EU 9710

Godiva Pump
Pump cap: 38 L/S
Tank cap: 900 L
Medium Pump

Another example of this popular appliance. It was still in service at Matata in the Bay of Plenty in 1993 and kept in almost new condition.

1967 BEDFORD KELC3
Reg. DO 8319

1967 BEDFORD KELC3
Reg. DN 7739

Coventry Climax Pump
Pump cap: 38 L/S
Tank cap: 900 L
Medium Pump
Body mfct: Wormalds

A crew cab Bedford with a purpose-built locker section, this was stationed at Bulls and is pictured here at Wanganui awaiting disposal.

1967 BEDFORD TK
Reg. DO 8340

Godiva Pump
Pump cap: 38 L/S
Tank cap: 900 L

This was the main pump at Okaihau in Northland but was found unsuitable due to lack of crew space in the cab; extra crew were forced to ride on the back. While it was away for repairs, the brigade borrowed an enclosed-cab relief pump and refused to return it afterwards. Following a number of meetings and much publicity, the NZFS finally agreed to a new appliance. This was purchased by the Kaipara District Council and is operated by the Ararua Fire Party.

Purchased by Southbridge Volunteer Fire Brigade in 1968 for £5188 with money given by the townspeople, this is now with the Kirwee Volunteer Fire Brigade in the South Island. Note the unusual scroll work above the front grille.

1968 BEDFORD KEL
Reg. DT 9147

One of three appliances based on the Hino chassis but fitted with a forestry style rear locker section (two others were at Sheffield and Hororata in the South Island), this one was stationed at Mapua. It has since been replaced.

1975 HINO KL340
Reg. AA 6777

1961 LANDROVER SERIES 2
Reg. DT 5367

KSB Pump
Pump cap: 20 L/S
Tank cap: 180 L
Light Pump

Stationed at Franz Josef in the South Island, this Landrover carried the suction coiled around the front bonnet and (usually) a standpipe on the brackets on the front bumper.

1964 LANDROVER FT6
Reg. AH 9891

Amag Pump
Pump cap: 29 L/S
Body mfct: Carmichael
Light Pump

Originally purchased by the Temuka Volunteer Fire Brigade, this was stationed there for most of its life until finally being transferred to Brooklands in Christchurch. It was built by Carmichaels Ltd in England and imported to New Zealand by Utilities New Zealand Ltd. It is one of about six of its type known to have been brought out to New Zealand and is now with the Fire Service Historical Society in Christchurch.

Commissioned as the number one pump at Auckland Headquarters on 17 June 1968, this was the first International to go into service in New Zealand. After serving at Headquarters, Ponsonby, Pt Chevalier, Manurewa, Titirangi and Mangere, it was downgraded to a relief pump in 1980.

1968 INTERNATIONAL AAC0183
Reg. DG 1462

A style often referred to the 'butter box', this open crew cab appliance was once stationed at the Stratford Volunteer Fire Brigade. It was the first of its type of pump to go into service with the volunteer fire brigades. The first-aid reel is directly behind the cab above the pump panel. It was downgraded to a reserve pump at New Plymouth region before being disposed of and purchased by the Tasman District Council.

1971 INTERNATIONAL C1800
Reg. FT 1971

1972 INTERNATIONAL 1820
Reg. GC 9099

Coventry Climax Pump
Pump cap: 38 L/S
Tank cap: 900 L
Medium Pump
Body mfct: Wormalds

Described as a rural pump, this appliance was located at Waihi Beach. It is understood that only two or three of these open-style cab appliances were built. It is now a relief pump at Tauranga, previously being stationed at Thames. It has an unusual Wormald body mounted on an International chassis.

1975 INTERNATIONAL ACCO
Reg. HG 5025

Darley 750 Pump
Pump cap: 57 L/S
Tank cap: 1350 L
Heavy Pump
Body mfct: Apex

Photographed at Edgecumbe in the Bay of Plenty, this appliance is the second pump for the brigade. It is another version of the 'butter box' International.

Darley Pump
Pump cap: 55 L/S
Tank cap: 1350 L
Heavy Pump
Body mfct: Apex

This was stationed at Te Aroha Volunteer Fire Brigade and replaced their Ford D600 which had been transferred to Raglan Volunteer Fire Brigade. The first-aid reel is in the centre locker instead of the usual roof-mounted version.

1975 INTERNATIONAL 1810A
Reg. HI 6817

Darley Pump
Pump cap: 55 L/S
Tank cap: 1350 L
Heavy Pump
Body mfct: Mills Tui

One of the earlier models of the International, this style appeared around 1975 and lasted until 1988 with minor body alterations. This appliance was stationed at Huntly Volunteer Fire Brigade.

1975 INTERNATIONAL ACCO 1810
Reg. HP 976

1969 INTERNATIONAL C1800
Reg. EF 2878

Darley Pump
Pump cap: 55 L/S
Tank cap: 900 L
Medium Pump

The first International-Darley appliance 'Fleet 60', as the pump was known. It introduced into the New Zealand fire-fighting scene the multi-stage American fire pump, the first use of high pressure fog hose reels and the electronic siren. The appliance went into service with the Auckland Metropolitan Fire Board in October 1969, and served at Headquarters, Otahuhu, Pt Chevalier and Ellerslie. During its 19 years of service it travelled some 300,000 miles (500,000 km). It was given to MOTAT and has since been restored.

1973 INTERNATIONAL F1800
Reg. GD 1709

Dennis No. 2 Pump
Pump cap: 38 L/S
Tank cap: 900 L
Medium Pump
Body mfct: Wormald Hale

The unusual cab design makes this appliance quite unique. Built in New Zealand by Wormald-Hale, it has spent all of its service life in Waipu, just south of Whangarei.

Waterous Pump
Pump cap: 55 L/S
Tank cap: 1350 L/S
Body mfct: Wormald-Hale
Heavy Pump

One of a pair built just prior to the New Zealand Fire Service change-over, this appliance is stationed at Maungaturoto. Its sister appliance is at Ruakaka, likewise in the Whangarei area. They were specially designed and built by Wormald-Hale with a spacious cab and locker area.

1975 INTERNATIONAL C1800
Reg. HM 2828

Gwynne Pump
Tank cap: 900 L
Body mfct: Carmichael, UK

The only one of its kind to be imported into New Zealand, by the Tauranga United Fire Board, this was fitted with a wheeled escape and later converted to a foam tender. In 1980 it was transferred as a pump to Paeroa Volunteer Fire Brigade and was finally withdrawn from service in 1988. It is now in Christchurch with the Fire Service Historical Society and the wheeled escape is in the Tauranga Historic Village.

1964 LEYLAND ALBION
Reg. EK 2348

1971 FORD D500
Reg. FW 6024

Dennis No. 2 Pump
Pump cap: 38 L/S
Tank cap: 675 L
Medium Pump
Body mfct: G.N. Hale

Built on a D500 chassis with an open-back crew cab, this appliance is now in service at Ikamatua, on the West Coast of the South Island. It was originally stationed at Eastbourne and when it left there for the South Island it had only 5500 miles (8900 km) on the clock.

1972 NISSAN TOILER
Reg. FP 1989

Dennis No. 2 Pump
Pump cap: 38 L/S
Tank cap: 1350 L

This unusual appliance was located at Omarama in service with the volunteer fire brigade there. It was downgraded to a training pump and sold in the early 1990s to the Queeenstown Lakes District Council.

This appliance has the polished metal body and locker section, a folding door to the crew cab and twin beacons. Carrying the old fleet number F23, it was stationed at the Cust Volunteer Fire Brigade and was still in service in early 1991.

1964 COMMER VAKS
Reg. DV 3450

A similar appliance to the one above, this Commer Vaks was stationed at the Springfield Volunteer Fire Brigade. The old fire crest on the door indicates that the photograph was taken prior to 1975 and the introduction of the 'Phoenix' crest. It has the hinged crew cab door and was one of the few that had the body painted. It has since been relocated to Fox Glacier Volunteer Fire Brigade.

1965 COMMER VAKS
Reg. EU 5967

1967 FORD D600
Reg. DP 2320

Dennis Pump
Pump cap: 58 L/S
Tank cap: 900 L
Medium Pump

Stationed at Invercargill, this Ford appliance has an unusual locker configuration with the pump mounted in the middle of the lockers. It is now a relief pump at Invercargill.

1967 FORD D600
Reg. CV 7628

Gwynne Pump
Pump cap: 38 L/S
Tank cap: 1350 L
Medium Pump

Formerly stationed at Omokoroa just north of Tauranga, this Ford is now the area relief pump at Tauranga. When photographed it still had the 'rubber duck' fire service emblem on the door.

Coventry Climax Pump
Pump cap: 30 L/S
Tank cap: 450 L
Body mfct: Wormalds

An earlier model of the Ford series, here photographed in Wellington, this has the lift-up style locker system, enclosed crew cab and the beacon fitted on the front centre of the roof. It is said to have been converted at one time to a foam tender for Petone. It has since been disposed of.

1968 FORD D600
Reg. EB 1574

Gwynne Pump
Pump cap: 38 L/S
Tank cap: 1350 L
Medium Pump
Body mfct: Wormalds

This appliance was stationed at Wairoa and at Ruatoria in the Gisborne area and then became an area spare. It has been fitted with a spare wheel on the front bumper and returned to service at the Ruatoria Volunteer Fire Brigade in 1993.

1969 FORD D600
Reg. FJ 9658

1969 FORD D600
Reg. ED 8525

Coventry Pump
Pump cap: 38 L/S
Tank cap: 1350 L
Medium Pump

An early model built on the Ford D600 chassis, this one has the folding doors on the rear crew cab. It was stationed at the Seddon Volunteer Fire Brigade.

1970 FORD D600
Reg. DU 8644

Godiva Pump
Pump cap: 38 L/S
Tank cap: 900 L
Medium Pump

This Ford model became the mainstay of a large number of volunteer fire brigades throughout New Zealand in the 1970s. This one was stationed at Oxford near Christchurch. It has the twin first-aid reels plus the open crew cab which undoubtedly makes things a little cold for crew on frosty nights.

Originally in service with the Lincoln Volunteer Fire Brigade, this appliance has now been relocated to Dunsandel.

1971 FORD D1013
Reg. GF 2460

Stationed on the West Coast of Northland at the Omapere Volunteer Fire Brigade, just past Opononi, this version of the Ford had hinged doors on the crew cab and the first-aid reels fitted in the lower section of the lockers behind the rear wheels. It has since been replaced by a Hino.

1971 FORD D600
Reg. FV 9447

1971 FORD D600
Reg. FW 4556

Coventry Climax Pump
Pump cap: 38 L/S
Tank cap: 900 L
Medium Pump
Body mfct: Wormalds

Photographed at the Fire Service workshops in Christchurch, this appliance was on its way south to Governor's Bay, and had just arrived from Wellsford. It has an unusual amount of white paint on the body and what appears to be a lower rear locker section.

1972 FORD D600
Reg. GB 5863

Dennis No. 2 Pump
Pump cap: 38 L/S
Tank cap: 900 L
Medium Pump

This appliance was based at Owhango, situated on the main highway through Tongariro National Park, for the volunteer fire brigade there. The members had a zip-up awning fitted to the open crew cab to keep out the winter cold. It also has the old-style lift-up locker doors. It has now been bought by the Far North District Council for the Horeke Fire Party.

Godiva Pump
Pump cap: 38 L/S
Tank cap: 450 L
Medium Pump

The unpainted locker
section on this vehicle
and the open crew cab
represent some of the
variations found on this
very common appliance
type. It was stationed for
a time at the Akaroa
Volunteer Fire Brigade
and has since been
relocated to Greymouth.

1972 FORD D600
Reg. FS 9018

Coventry Climax Pump
Pump cap: 38 L/S
Tank cap: 900 L
Medium Pump
Body mfct: Wormalds

One of the last group of
this style of appliance,
this shows a more
streamlined body with a
higher locker section.
This one is stationed at
Okaiawa Volunteer Fire
Brigade, just out of New
Plymouth.

1974 FORD D1013
Reg. HZ 6785

1972 TOYOTA LANDCRUISER

Reg. FV 6193

Gywnne Pump
Pump cap: 30 L/S
Tank cap: 450 L
Light Pump

This appliance was stationed on Waiheke Island for a time and was later sold to the General Foods Industrial Brigade. It was sold again when this brigade was disbanded.

1974 TOYOTA LANDCRUISER

Reg. GH 3356

Dennis Pump
Pump cap: 38 L/S
Tank cap: 680 L
Light Pump

Stationed with the Russell Volunteer Fire Brigade, this was one of three of its type in the Northland area. The others were at Kamo near Whangarei and at Okaihau. These appliances were ideal for their off-road abilities, especially in fighting scrub fires. All three have since been disposed of.

Dennis Pump

One of three of this type of appliance in service with brigades in Region 1B in the north, this Landcruiser was stationed at Okaihau. The others were at Russell and Kamo. All three have since been sold to the Far North District Council for use by fire parties.

1975 TOYOTA LANDCRUISER
Reg. HH 8235

Tohatsu Pump
Pump cap: 20 L/S
Tank cap: 450 L
Light Pump

Previously stationed at Tahuna and Murupara, this four-wheel-drive appliance is now on the Chatham Islands where the brigade is often referred to as New Zealand's most isolated. This is their second appliance, with a 1981 International being their number one pump. It is photographed here in the rear yard of Rotorua Station.

1975 TOYOTA LANDCRUISER
Reg. HC 6523

New Zealand Fire Service 1976–

The New Zealand Fire Service Commission (NZFS) took over the total control of former local government fire brigades and fire boards from 1 April 1976. This replaced the Fire Service Council and nationalised all of the local authority based services.

There were 277 Fire Districts, 26 served mainly by permanent fire-fighters and the remaining 251 entirely by volunteers, as well as 33 Auxiliary Districts.

The assessment of the inherited assortment of vehicles and buildings was a mammoth task. Although some of the larger, more modern metropolitan stations were well stocked and designed, many stations left a lot to be desired.

Many city stations were old but they were strong and serviceable. The rural stations were rather different. Many had begun life as a shed to house the portable pump, then enlarged to hold an appliance. Often they were older buildings altered to suit or built from less permanent material.

The efforts of the NZFS to update these buildings over recent years have included the use of modular type buildings of both single and double bay designs, which are both functional and soundly constructed. Local brigades can enhance these by 'self-funded' additions such as social rooms.

The appliances themselves were a really mixed bag. In 1976 it was reported that the NZFS had inherited 1879 permanent fire-fighters, 6434 volunteer fire-fighters, 470 assorted appliances (self-propelled pumps), 430 portable pumps and 105 trailer pumps. Although the metropolitan areas managed to achieve some standardisation, the balance of the fleet was very mixed. It ranged from 'hand-me-downs' from the city brigades to pre-war open appliances. As many vehicles were still on order to former fire boards, the replacement of appliances seemed slow at first.

After the initial review of the entire fleet, some relocations took place to meet national requirements. The old system of passing down older city appliances to smaller brigades soon started to change as the commission was able to match and juggle the appliances to suit local needs.

In the past, the brigades and authorities had relied on what was available or appliances were assembled by Mills Tui or Wormalds to a broad set of standards. Individual brigades did not have the technical knowledge to design equipment suitable for local requirements.

The National Fire Service, with a fleet of 800 plus self-propelled pumps and other specialist equipment, was able to put its research and technology into designing a range of specific vehicles for individual areas and needs.

The commission has created a range of classifications for the different types of appliances. The most common specifications are:

6/1	Light Pump	6/11	Hydraulic Elevating Monitor
6/2	Standard Pump	6/13	Rescue Tender — Light
6/3	City Pump	6/14	Rescue Tender — Heavy
6/7	Light Four-wheel Drive	6/15	Hose Layer
6/8	Standard Rural	6/16	Foam Tender
6/9	Hydraulic Elevating Platform	6/21	Water Tender
6/10	Turntable Ladder	6/24	Emergency Tender — Heavy

During the 1980s, the NZFS began matching vehicle resources to the fire district size. This was done by defining the operational requirements and then matching one of the three basic pumps to meet them.

Three standard pumping appliances are built in New Zealand to meet operational requirements. Each type is built on a chassis and crew cab unit which meets NZFS regulation ECE29 or equivalent. The three pump types and their requirements are:

LIGHT PUMP:

Weight not to exceed 10 tonnes	Diesel engine rated at 160–180 bhp
Acceleration to 65 km/h in 30 seconds	Road speed of 100 km/h
Pump to deliver: 30 L/S at 1000 kPa	2000-litre water tank
6 L/S at 1800 kPa	

MEDIUM PUMP:

Weight not to exceed 12 tonnes	Diesel engine rated at 180–220 bhp
Acceleration to 65 km/h in 25 seconds	Road speed of 110 km/h
Pump to deliver: 55 L/S at 700 kPa	1350-litre water tank
30 L/S at 1000 kPa	
8 L/S at 1800 kPa	

HEAVY PUMP:

Weight not to exceed 13 tonnes	Diesel engine rated at 220–260 bhp
Acceleration to 65 km/h in 18 seconds	Road speed of 110 km/h
Pump to deliver: 55 L/S at 700 kPa	1350-litre water tank
30 L/S at 1000 kPa	
8 L/S at 3500 kPa	

Simultaneous operation of high pressure hose reels and low pressure deliveries to set performance standards.

The first of these, built by Mills Tui in 1987, included the new fire service crew cab which was on a Hino FD chassis. By July 1991 they had built 50 of these appliances.

The NZFS has now established period supply contracts by competitive tender with selected manufacturers. This has seen the importation of cabs and chassis from Dennis, International, Hino, Mitsubishi and Scania, with Mills and Tui of Rotorua and Lowes Industries of Wellington assembling them and fitting the bodies to the NZFS specifications. This has allowed the NZFS to plan its replacement programme more efficiently.

Although the commission is still moving some of the older more serviceable vehicles to smaller units, the new replacement policy will soon end this. In future the life expectancy of each type of appliance will receive greater consideration. Since the smaller 'type one' pumps have a higher work load, they will be replaced sooner than the bigger 'type two' or 'type three' machines.

The other factor is that the actual mileage that each appliance travels is usually not very high. Many an appliance sold after 20 years of service has travelled less than 24,000 km (15,000 miles). The strain on the motors of the appliances is much greater than with a truck of similar size and weight. The fire appliance is checked at least once or twice a week, as are all the portable pumps and equipment, to ensure that they are operational immediately as and when required.

Most of the appliances have a built-in warmer system to ensure that the motor is warm and easy to start when required. This helps reduce the wear on the motors. Once the brigade is alerted for a call-out, they require the appliance to be able to respond. In the early days nothing was more embarrassing than the sight of burly fire-fighters, fully kitted out, push-starting their appliances!

1982 ISUZU Model SBR
Reg. KY 4037

Darley HM 350 Pump
Pump cap: 30 L/S
Tank cap: 950 L
Body mfct: Mills Tui
6/1 Light Pump

Powered by an Isuzu diesel motor, this machine is the only one of its type in NZFS service. It was originally built for an industrial brigade and is now stationed at Kawhia.

1979 FUSO Model FK 102
Reg. JL 9424

Darley HM 350 Pump
Pump cap: 23 L/S
Tank cap: 1350 L
Body mfct: Mills Tui
6/1 Light Pump

The only one of its type in New Zealand, this Fuso has an open rear crew cab and a small locker section. It was photographed in service at Riversdale.

Waterous Pump
Pump cap: 55 L/S
Tank cap: 1364 L
6/3 Heavy Pump

This Ford, powered by a
Cummings V8 504 diesel
and based at Patumahoe,
has an enclosed rear
crew cab.

1976 FORD D1618
Reg. IN 9557

Waterous Pump
Pump cap: 79 L/S
Tank cap: 1270 L
Body mfct: Wormalds
6/3 Heavy Pump

This Christchurch-based
appliance is similar to
that in the photograph
above. The main
difference is that the rear
locker section is in
polished aluminium with
a built-in ladder for access
to the rear ladders.

1976 FORD Model D1618
Reg. IN 9556

1977 FORD Model D1011
Reg. IE 3843

Darley Pump
Pump cap: 25 L/S
Tank cap: 1800 L
Body mfct: Mills Tui
6/1 Light Pump

Said to be the first Ford appliance built by Mills Tui of Rotorua for the NZFS, and powered by a Ford petrol motor, this appliance has the open rear crew cab. It is stationed at Mangakino.

1991 IVECO Cargo 1218
Reg. PO 3077

Waterous Pump
Pump cap: 38 L/S
Tank cap: 1350 L
Body mfct: Lowe Industries
6/2 Medium Pump

One of the first batch of 21 of this type of appliance built in 1991 as medium pumps, this Iveco is powered by a B6TA Cummings diesel motor, and is stationed at Kaitaia in the Far North.

Darley SEH500 Pump
Pump cap: 41 L/S
Tank cap: 1800 L
Body mfct: Mills Tui
6/2 Medium Pump

An open crew cab version on a Commer chassis and powered by a Chrysler petrol motor, this appliance operates in the Northland town of Kaikohe.

1977 COMMER Model RG11
Reg. IN 9880

Darley SEH500 Pump
Pump cap: 30 L/S
Tank cap: 1800 L
Body mfct: Mills Tui
6/2 Medium Pump

Powered by a Perkins diesel motor, this is stationed at Southbridge in Christchurch. This model has the enclosed rear crew cab.

1978 COMMER Model RG13
Reg. IS 2312

1978 COMMER Model G1311
Reg. JD 8851

Waterous Pump
Pump cap: 45 L/S
Tank cap: 1800 L
Body mfct: Wormalds
6/2 Medium Pump

This appliance, powered by a Perkins diesel motor, was stationed at Middlemarch in Otago.

1980 DODGE Model R1513
Reg. JD 7619

Darley SEH500 Pump
Pump cap: 47 L/S
Tank cap: 1800 L
Body mfct: Mills Tui
6/2 Medium Pump

An open crew cab appliance powered by a Perkins diesel motor, this is stationed at Paekakariki, north of Wellington.

1978 BEDFORD Model MK65
Reg. JD 9048

Waterous Pump
Pump cap: 28 L/S
Tank cap: 1200 L
Body mfct: Wormalds
6/7 4x4 Pump

Some eighteen of this type of appliance were built for the NZFS in 1978 and allocated to a number of rural brigades. Built on a Bedford four-wheel-drive chassis, it is powered by a Bedford diesel motor. This one is stationed at Waitotara, south of New Plymouth.

1978 BEDFORD Model MK65
Reg. JD 9045

Waterous Pump
Pump cap: 33 L/S
Tank cap: 1450 L
Body mfct: Wormalds
6/7 4x4 Pump

On the same MK65 Bedford chassis as the previous appliance but with a larger rear locker section, the pump panel and first aid reel have been located at the rear. It also has a larger water tank. It is stationed at Ohakune, south of Mt Ruapehu.

1979 BEDFORD Model MK65
Reg. JK 7031

Darley Pump
Pump cap: 23 L/S
Tank cap: 1360 L
Body mcft: Mills Tui
6/7 4x4 Pump

Again this open-cab appliance has a Bedford four-wheel-drive chassis and has a mid-mounted pump. It is stationed at Raetihi near the southern portion of Tongariro National Park.

1978 MACK Model CF 685
Reg. JR 426

Waterous CMYB Pump
Pump cap: 78 L/S
Tank cap: 1550 L
Body mfct: Wormalds
6/3 Heavy Pump

Resembling an American appliance, this is powered by a Mack diesel motor and, when photographed, was a relief pump for the Wellington region.

1983 DENNIS Model DF133
Reg. LH 1336

Darley SEH 750 Pump
Pump cap: 61 L/S
Tank cap: 1550 L
Body mfct: Mills Tui
6/3 Heavy Pump

Photographed at Johnsonville in the Wellington region, this was the only example of cab and chassis model DF133 that the NZFS put into service — they opted for model DF135 instead.

1985 DENNIS Model DF135
Reg. MH 8865

Darley SEH 750 Pump
Pump cap: 55 L/S
Tank cap: 1350 L
Body mfct: Mills Tui
6/3 Heavy Pump

One of 20 of this appliance model built for the NZFS in 1985-86, this is powered by a Perkins diesel motor and used mainly in urban areas. This one was stationed at Henderson in West Auckland as the volunteer brigade's pump.

1976 CHEVROLET Model C30
Reg. HN 1534

Darley HM 350 Pump
Pump cap: 30 L/S
Tank cap: 900 L
6/1 Light Pump

Stationed at Benneydale
(ex-Te Aroha), this has a
higher rear locker section
than the appliance below
and the first-aid is fitted
into the lower rear
section.

1978 CHEVROLET Model C30
Reg. IT 7395

Hale CBP 3 Pump
Pump cap: 15 L/S
Tank cap: 485 L
Body mfct: Wormalds
6/1 Light Pump

Powered by a Chevrolet
petrol motor, this
appliance is pictured at
Akaroa as their second
pump. Only two of the
type were ever made.

Powered by a Dodge petrol motor, this appliance was stationed at Te Puia Springs.

1977 DODGE Model D5N
Reg. HD 2808

This appliance is a sister to that in the above photo, but painted in an unusual colour scheme. It is built on a Dodge light chassis with enclosed crew cab. The photograph was taken at Kinloch.

1977 DODGE Model D5N
Reg. HD 2810

1976 HINO Model KR 320E
Reg. HZ 2967

Dennis #2 Pump
Pump cap: 29 L/S
Tank cap: 1300 L
6/1 Light Pump

This unusual Hino, powered by a Hino EH 100 diesel motor, is stationed at Coalgate in Canterbury. It has a single cab and a smaller rear locker section.

1977 HINO Model KR 320
Reg. IR 4948

Waterous Pump
Pump cap: 57 L/S
Tank cap: 1250 L
6/2 Medium Pump

An early model Hino, powered by a diesel motor and stationed at Kawakawa in Northland. Only about five of this model went into service as pumps.

This is the only one of its type known to be in operation, with a similar rear locker section to the D200. It is stationed at Tangimoana.

1976 INTERNATIONAL Model D1310
Reg. IE 6431

Only about six of this type of pump were ever built. Based on an ambulance chassis, most have since been recommissioned as emergency tenders. Originally they had only single rear wheels but dual wheels were fitted for better stability. The photo was taken at Ahipara in the Far North.

1976 INTERNATIONAL Model D200
Reg. IJ 1950

1976 INTERNATIONAL Model 1710A
Reg. IJ 1987

Darley HM 500 Pump
Pump cap: 32 L/S
Tank cap: 1350 L
Body mfct: Mills Tui
6/2 Medium Pump

The missing stripe on the rear crew cab door and the placement of the NZFS crest makes this appliance different. It is stationed at Ashburton and powered by a petrol motor. Only seven of this model were built during 1976.

1981 INTERNATIONAL Model 510A
Reg. KE 7222

Darley HM 350 Pump
Pump cap: 26 L/S
Tank cap: 1350 L
Body mfct: Mills Tui
6/1 Light Pump

Only three of this type of appliance are known to have been in service with the NZFS. They are built with a 610A cab and chassis but have four-wheel-drive. The open rear crew cab and smaller rear locker section makes this appliance unique. Stationed at Okato in the New Plymouth area, it also has a front-mounted power winch.

1982 INTERNATIONAL Model 1950C

Reg. JZ 6233

Darley SEH 740 Pump
Pump cap: 58 L/S
Tank cap: 1350 L
Body mfct: Mills Tui
6/3 Heavy Pump

A Cummings diesel powers this appliance which is stationed at Stoke in the South Island. The rear section has a greater number of lockers than usual as well as extra ladders. Thirty-eight of these appliances were built between 1981 and 1984.

1982 INTERNATIONAL Model 1810C

Reg. KR 4565

Waterous Pump
Pump cap: 44 L/S
Tank cap: 1350 L
6/2 Medium Pump
Body mfct: Wormalds

Stationed at Kaiapoi in the South Island, this International is powered by a petrol motor and, unlike most, has the first-aid reel fitted in the lower section of the rear lockers (refer page 110). A total of 57 of this model were built and put into service between 1981 and 1983. Some were constructed by Wormalds with the waterous pumps but the majority were built by Mills Tui and fitted with the popular Darley SEH 500 pumps.

1977 INTERNATIONAL Model 1910A
Reg. IR 9031

Tank cap: 1800 L
Darley Pump
Body mfct: Mills Tui
6/2 Medium Pump

Only six of this early model International were built. One was a hoselayer based in Wellington, and the other five were pumps. They were all powered by International petrol motors. Three were built by Wormalds and the others were built by Mills Tui. This one was stationed at Manly, an auxiliary to the Silverdale Brigade.

1981 INTERNATIONAL Model 610A
Reg. KE 7297

Darley HM 350 Pump
Pump cap: 30 L/S
Tank cap: 1300 L
Body mfct: Mills Tui
6/1 Light Pump

An open cab style International powered by a petrol motor, this was photographed on station at Kaiwaka in Northland. It has since been transferred to Ahipara in the Far North. A total of 12 of this type of appliance were built in the period 1981–1982.

A petrol-powered International, used widely by many of the volunteer brigades, the rear locker section appeared in a number of variations. This is stationed at Sumner near Christchurch.

1981 INTERNATIONAL Model 1810C
Reg. KK 9182

One of the last of the Internationals, fitted with a Perkins diesel motor, this was stationed at Brooklands in Christchurch. Only 16 of this model were built, and all were fitted with Perkins Phaser diesel motors and Rosenbauer pumps.

1988 INTERNATIONAL Model 1850D
Reg. OE 9038

1981 JEEP Model J20
Reg. JZ 5575

Darley HM 350 Pump
Pump cap: 26 L/S
Tank cap: 550 L
Body mfct: Mills Tui
6/7 Light pump (4x4)

Eighteen of these were put into service; they were ideal for country areas where access was often difficult. This one was stationed at Te Awamutu.

1987 VOLVO Model B6FA
Reg. ND 8790

Darley SEH 750 Pump
Pump cap: 40 L/S
Tank cap: 1350 L
Body mfct: Mills Tui
6/2 Medium pump

One of six of these appliances built in 1987, this has a similar cab to the Mitsubishi and is powered by a Volvo diesel motor. It is now stationed at Waipawa in Hawke's Bay.

Darley RM500 Pump
Pump cap: 40 L/S
Tank cap: 2700 L
Body mfct: Mills Tui
6/1 Light Pump

Stationed at Bulls, this is one of the earlier Hinos that were manufactured between 1983 and 1985 and placed into service with several Volunteer Brigades. From 1983 to 1986 a total of 29 of this model were produced. They were all powered by a Hino diesel motor and 10 were fitted with the Darley RM 500 pumps while the balance were fitted with the waterous CP3 pumps.

1984 HINO Model FF173
Reg. LP 4131

Darley HM500 Pump
Pump cap: 30 L/S
Tank cap: 2000 L
Body mfct: Mills Tui
6/1 Light Pump

Based at the Northland settlement of Okaihau, this was one of the first of the new smaller appliances to be seen north of Auckland. It replaced their old open-back style TK Bedford as a result of protest by the local brigade. Some 52 of this popular appliance were constructed by Mills Tui between 1987 and 1990. They were powered by a Hino diesel motor and fitted with Darley HM 500 pumps.

1987 HINO Model FD162LA
Reg. NR 8404

1989 HINO Model FH222
Reg. OF 5623

Rosenbauer Pump
Pump cap: 38 L/S
Tank cap: 1550 L
6/2 Medium Pump

Powered by a Hino EM 100 diesel motor, this was one of 19 built in 1989 on the bigger Hino chassis to NZFS 6/2 specifications. The photograph was taken at Christchurch Training. The appliance is now stationed at Lake Tekapo.

1990 HINO Model FD1017
Reg. PN 1268

Darley HM500 Pump
Pump cap: 30 L/S
Tank cap: 2000 L
Body mfct: Mills Tui
6/2 Medium Pump

Based at Waikari in Canterbury, this appliance is powered by a Hino diesel motor and carries a set of hose protection ramps on the front bumper. One of seven of this model built during 1990.

Darley SEH 500 Pump
Pump cap: 38 L/S
Tank cap: 1350 L
Body mfct: Mills Tui
6/2 Medium Pump

Twenty of these appliances were put into service in 1985. This one has the twin first-aid reels above the pump panels behind the cab. It is stationed at New Brighton, Christchurch.

1985 MITSUBISHI Model FP418JR
Reg. MC 3851

1987 MITSUBISHI Model FP418JSRFB2
Reg. NK 5410

Darley Pump
Pump cap: 55 L/S
Tank cap: 1450 L

Some 37 of this model of appliance were built in 1987-1988 for the NZFS by Mills Tui. All were powered by Mitsubishi diesel motors and they were all fitted with the Darley SECH 1000 pump. Fifteen of them were converted to Pump/ Rescue appliances like the one shown, which is based at Wanganui.

1989 MITSUBISHI Model FP418
Reg. OA 3874

Darley SECH 1000 Pump
Pump cap: 55 L/S
Tank cap: 1350 L
Body mfct: Mills Tui
6/3 Heavy Pump

This is a later model Mitsubishi with a heavier pump used for training in Christchurch. It is now stationed at Addington in Christchurch and has a larger number of rear lockers than usual.

1992 MITSUBISHI Model FK160F
Reg. RR 4002

Darley HM 500 Pump
Pump cap: 30 L/S
Tank cap: 2000 L
Body mfct: Mills Tui
6/1 Light pump

The new cab style of the Mitsubishi light pump, this one is powered by a Mitsubishi diesel and is stationed at Te Kopuru, an auxiliary of Dargaville.

Photographed at the Auckland NZFS workshop prior to being commissioned at Otahuhu station, this is one of the first of the new Mitsubishi heavy pump appliances.

1992 MITSUBISHI Model FP 41J
Reg. SC 1054

The Scania cab and chassis come from one of the few manufacturers worldwide to satisfy the new regulations that govern the standards and dimensions of the double fire cab. The suspension and brakes have been especially developed by Scania for the operational needs of fire appliances. This vehicle is stationed at Hamilton.

1989 SCANIA Model G93M
Reg. OS 3601

Specialist appliances

While the primary purpose of the fire appliance at first was to fight fires using the basic pump and hose, the trend towards multi-storied buildings required tall ladders for both rescue and fire-fighting.

The earliest ladders were the horse-drawn escape ladders, some capable of reaching up to 20 metres. These were replaced by the motorised pump escape vehicles and then the larger Turn-table Ladder vehicles (TTL). The 1924 Tillings Stevens TTL was used at the Ballantynes fire in Christchurch in 1947. This type was replaced by Leylands and Dennis-Metz TTLs which often featured the four-section Metz telescopic turn-table ladders that could reach up to 30 metres. These were later replaced by Ford Internationals and the Mack Aerialscope.

The development of high-rise office buildings also resulted in the need for such appliances as the Snorkels (HEP), Telesquirts and Telebooms (PHEM). The latter were pumping appliances equipped with special booms capable of delivering large volumes of water onto fires.

Many of the larger city areas also have Foam Tenders and Hose Layers, both capable of carrying quantities of waterway equipment and essential for fires involving flammable liquids.

At larger fires the Mobile Command Units and BA tenders (Breathing Apparatus) are essential, as are the Canteen Unit or Salvage Tender. With the widespread use of chemicals, the Fire Service has specialised HAZ-MAT (Hazardous Materials) units. These carry the necessary equipment and materials to dispose of dangerous and toxic chemicals.

As well as the fire-fighting role, fire brigades also become involved in rescue. They are called out to assist and free trapped victims at motor vehicle accidents as well as industrial accidents. Many brigades have purchased their own specialised equipment, including the 'Jaws of Life' and spreading and cutting tools. This is the hydraulic rescue equipment designed to cut and remove sections of vehicles to extract trapped victims. They also have a range of other equipment including air rescue bags designed to lift heavy objects off a victim. To carry this and other equipment, many brigades have their own Emergency Tenders (ET), ranging in size from the CF Bedford and Ford Transit up to more elaborate units. These vehicles and their equipment are often funded by a community or community groups who assist in obtaining and outfitting them. The larger city brigades have special heavy rescue tenders, sometimes with hydraulic lifting arms fitted to the rear.

As a number of rural areas are without reticulated water supply, some brigades operate their own Water Tanker. These are sometimes provided by the local district council. They ensure that plenty of water is available at the scene to assist in controlling larger fires.

Some brigades, because of their location, are also involved in cliff rescue. Recently, in the Far North, a surf rescue group was formed at Ninety Mile Beach.

THE CONSTRUCTION OF A FIRE ENGINE

The cab and chassis are delivered to the body builders. These are usually fully completed with radios and electronic equipment installed. The cab is usually painted and the beacons fitted. The first task is to install the pump and all the 'plumbing'; this includes the panels, gauges and the first-aid reels. Meanwhile the rear locker section is being fabricated. These rear sections are usually designed to fit around the water storage tank. Most of the construction is done in aluminium and sections welded together. Once the locker section is completed, it is fitted to the cab and chassis and final work carried out. Shown here is the manufacture of a light pump on an Isuzu SBR chassis. (Refer page 95.)

1979 BEDFORD CF
Reg. JS 2087

Popular with many brigades throughout New Zealand for MVA or rescue incidents, this type of appliance usually carries an assortment of equipment such as 'jaws of life', air bags, portable saws, lighting equipment and rescue lines. Many are owned or funded by local communities. This one is stationed at Ngaruawahia.

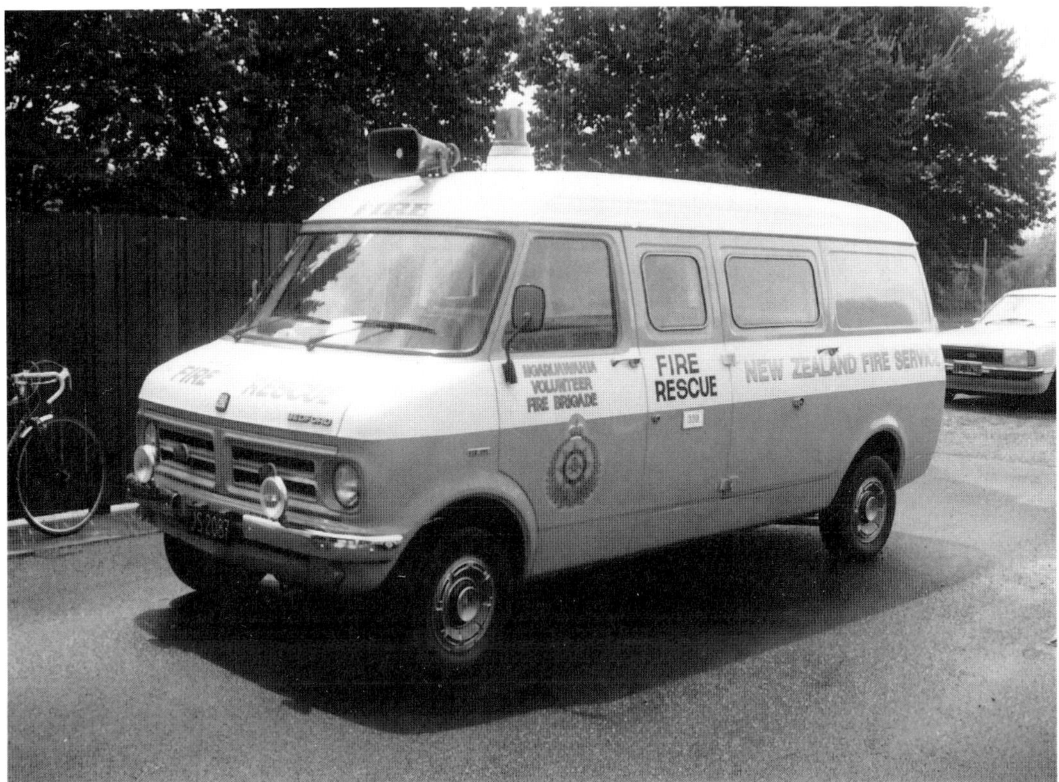

1982 BEDFORD CF
Reg. KW 8182

A larger version of the CF Bedford, this has a power winch fitted to the front to assist in rescues. Based at Maungaturoto, it attends a number of MVAs in the Brynderwyn Hills on State Highway 1 just south of Whangarei.

Stationed at the busy West Auckland station of Henderson, this emergency tender carries a full range of equipment including air equipment, generator and BA units, and can carry up to five personnel. It was originally a personnel carrier for the training division. The conversion was carried out by the members of the composite Henderson Brigade.

1987 DAIHATSU DELTA
Reg. NA 8775

This four-wheel-drive Toyota was purchased by the Paeroa Volunteer Fire Brigade and was converted to an ET, fitted out to carry all of their rescue equipment, including jaws of life and other cutting equipment.

TOYOTA LANDCRUISER
Reg. JF 6264

1967 DODGE AT4
Reg. DF 3877

A number of brigades used ex-ambulances as their rescue tenders. Most have since been disposed of in favour of Bedford vans or pump/rescue appliances. This one was stationed at Te Awamutu.

1984 BEDFORD CF
Reg. KT 9598

Stationed at Paihia, this former ambulance was converted to carry rescue equipment. It has since been decommissioned and rescue equipment is now carried on the No. 2 pump there. A first response ambulance unit has been formed in Paihia.

1985 TOYOYA HI-LUX
Reg. MF 4945

During the February 1985 floods, the Thames Volunteer Fire Brigade realised that it needed a four-wheel-drive emergency tender. After receiving donations from local business people for assistance given during the flood, the brigade launched a public appeal and within two months had raised almost $30,000 to purchase and fit out a special rescue tender with radio, jaws of life and cutting tools. This vehicle was dedicated in 1986 and can carry a team of five — an officer, driver and three crew.

1981 INTERNATIONAL ACCO 610A
Reg. KE 5044

This four-wheel-drive rescue tender was given to the Motueka Volunteer Fire Brigade by the Lions Club and people of Motueka. A special committee was set up to raise the funds, the Motueka Emergency Services Society Inc. or MESSI as it was known locally. The appliance is fitted with various rescue equipment and more is added periodically as funds become available.

1970 INTERNATIONAL C1300
Reg. JD 2991

Photographed outside the Tokoroa Fire Station, this appliance was originally a forest fire patrol unit. The body was built locally with the brigade paying for materials and the work provided by voluntary labour. The pump-electric clutch is driven from the fan belt. Its front-mounted winch is operated from a PTO from the gear box. It entered service in April 1979 and was disposed of in February 1991.

1984 DAIHATSU V54W
Reg. LI 2023

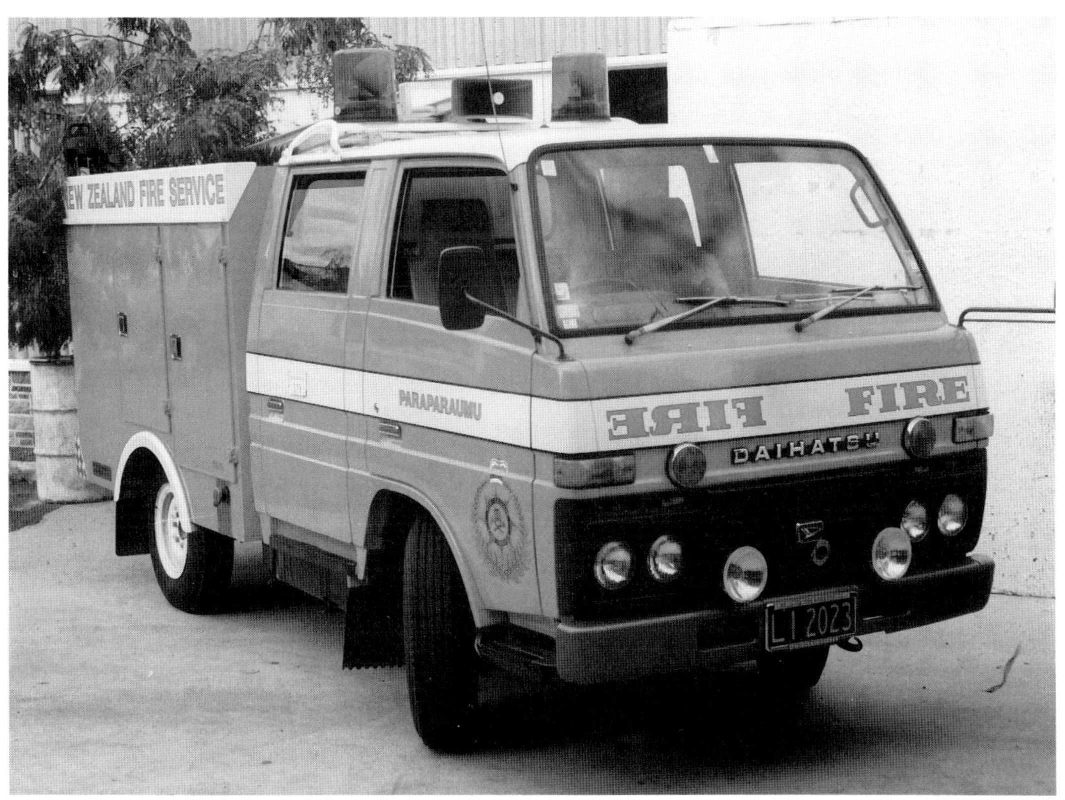

A larger version of an ET built on a Daihatsu crew cab chassis for the Paraparaumu Volunteer Fire Brigade. It was funded by the local community and given to the NZFS.

Built especially as a rescue tender with no water tank, this was based at Headquarters in Auckland before being transferred to Manukau. It has since been disposed of and is now in private ownership.

1971 INTERNATIONAL 1820
Reg. FP 1548

The appliance is shown at Christchurch where it was used for a period as a pump/rescue tender. It has unusual black and white checkered strips and it also had blue flashing beacons. More recently it has been recommissioned back to a pump.

1980 INTERNATIONAL 2150B
Reg. JR 600

1986 DENNIS DF135
Reg. MR 6877

Darley SEH750 Pump
Pump cap: 55 L/S
Tank cap: 1350 L
Pump/Rescue
6/23

The new concept within the Fire Service is to combine rescue equipment on a standard pump appliance. This one is stationed at Avondale Brigade in West Auckland. It is manned by permanent fire-fighters and responds to most incidents within the area.

1986 INTERNATIONAL 1950C
Reg. LH 934

Heavy Rescue Tender
6/24
Body mfct: Mills Tui

This appliance, stationed at Manukau, is the second heavy rescue tender in the Auckland area. It has a Palfinger crane mounted on the rear which has a maximum lift of 3750 kg, utilised in various rescue situations. It has also an air compressor mid-mounted and four air outlets plus the two air reels. The compressor supplies air for the various pneumatic rescue tools carried which include the jaws of life cutting tools and air bags. It also carries oxyacety-lene cutting equipment and a lighting plant.

Purchased and donated by the New Plymouth Jaycees, this appliance had a special body built and was supplied complete with rescue equipment. Based at the New Plymouth Fire Station and manned by permanent fire-fighters, it was normally turned out with one officer and one fire-fighter plus a pumping appliance and crew. It carried a large range of rescue equipment, including chemical suits, air equipment, lighting equipment, resuscitator, lines and first-aid equipment. It was decommissioned and sold in September 1989.

1970 INTERNATIONAL C1500
Reg. FM 4089

An unusual vehicle based with the Te Kuiti Volunteer Fire Brigade in the Hamilton region, this rescue tender is on a one-tonne Holden utility chassis with a rear locker section.

1976 HOLDEN
Reg. HW 7780

1970 AEC MONARCH
Reg. FK 3839

Command Unit
Body mfct: Chapmans, Nelson

This appliance was built on a Monarch chassis for the Dunedin Fire Brigade. It was built by Chapmans of Nelson to their design as a command unit. It was also known as a SECT (Salvage and Emergency Tender) and could be used as a mobile command post on all major incidents. It has been replaced with a new Mitsubishi appliance and is now at the Fire Service Historical Society in Christchurch.

1972/73 TOYOTA DYNA
Reg. HG 4587

Emergency Tender/ Command Unit

Built in Taupo in 1972 and commissioned as an emergency tender, this carried a range of rescue equipment and has a 5-kVa electrical generator. Transferred to Invercargill in 1976, it served there until 1982 when it was transferred to Palmerston North and used as a command unit and light generator. It was withdrawn from service and disposed of in 1987.

1976 INTERNATIONAL 1710A
Reg. IC 1355

Especially designed as a mobile command unit, this appliance responds to all major incidents in the Auckland area. It is set up as a mobile radio room, conference room and kitchen. A rear-mounted generator supplies power to the unit when it is stationary. It carries a number of portable field telephones and can be connected to the Telecom telephone system. It sports a chequerboard paint scheme and has a large array of aerials.

1988 MITSUBISHI
Reg. OM 3601

This is an example of the latest style of specialist appliance, with about four of these units having been built to date. Based at Lower Hutt, this one has the dual roles of a BA tender and a decontamination unit. It is fitted with an Atlas Copco BP3 compressor for refilling the BA units and carries 20 BA sets plus 20 spare cylinders, as well as full level-three decontamination equipment.

1934 FORD V8
Reg. EU 3542

Salvage Tender

This Ford, supplied to the Wellington Fire Board, was a fully enclosed appliance and an advanced machine for its time. Very popular with the members of the brigade, it continued in service until 1972. It carried a range of equipment including salvage sheets, mops, squeegees, buckets and sawdust. It is now at the Fire Service Historical Society in Christchurch.

1982 HINO FF173
Reg. KU 1469

Hazmat/Command 6/14

Located at Palmerston North, this specialist appliance is used for all major incidents, and carries all the necessary equipment for handling a chemical spill or incident. It is one of two of this style, the other is in Hawke's Bay.

1972 INTERNATIONAL C1800
Reg. GC 6339

This was purpose built for Hamilton in 1972 as a rescue unit. In 1988, with the arrival of a new pump/rescue appliance, it was converted to a Hazmat/Command Unit. Originally it had a winch fitted in the front bumper but this was removed in 1988. It is still in service with the Hamilton Brigade.

1983 FORD F300
Reg. KO 2598

A smaller version of a hazardous material appliance, this one is based at the Rotorua Station; there is a similar one at the Dunedin Station. It carries the equipment and neutralisers for chemical incidents.

1973 FORD D
Reg. GU 2577

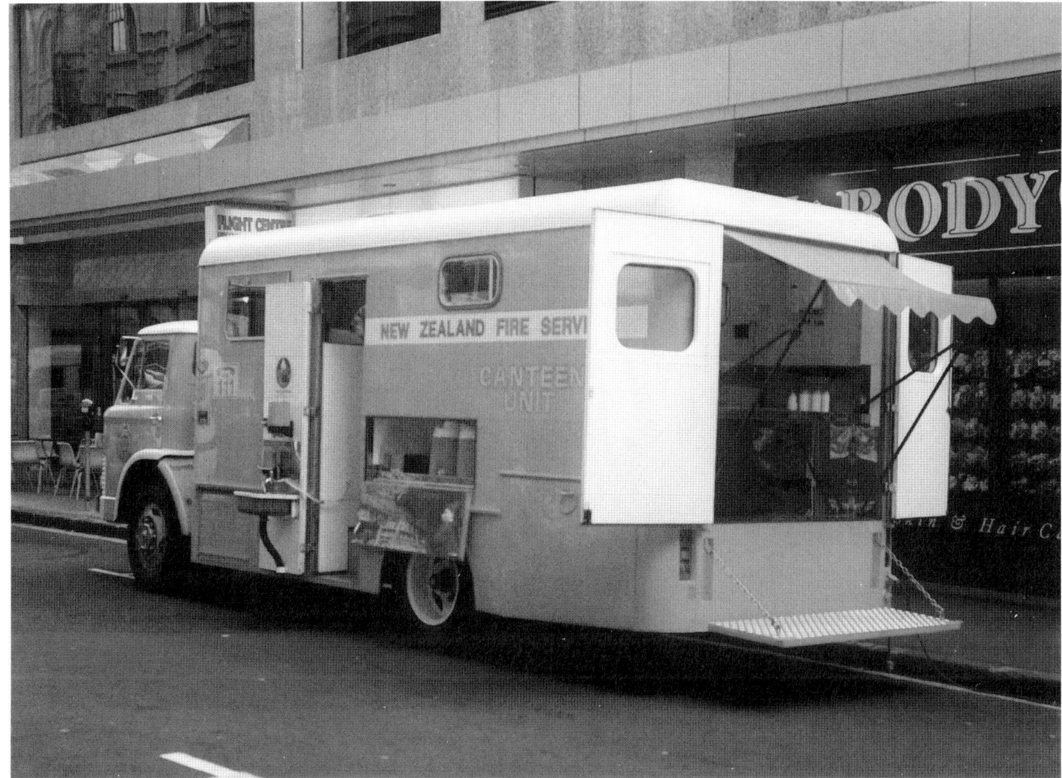

Canteen Unit

Initially this was built by the NZFS as a mobile training unit. It was converted and in 1981 recommissioned as a mobile canteen, complete with cooking and refrigeration facilities, and is used at all large incidents. It is stationed at Balmoral in Auckland.

1980 DODGE G1513
Reg. JX 9016

Darley Pump
Pump cap: 60 L/S
Tank cap: 1800 L foam
Foam Tender
6/16
Body mfct: Mills Tui

Built for the Auckland area, this appliance was commissioned early in 1981. It has a foam tank capacity of 1800 litres plus drum storage in the side lockers. Its 60 L/S pump can supply mechanical foam through side deliveries to foam branches or out of its roof-mounted foam monitor. It is specially designed for use at tank farms or other incidents involving flammable liquids where foam is required.

This tender provides on-site breathing apparatus facilities at large or serious incidents in the Auckland area. It has its own breathing apparatus compressor and can supply 83 cubic feet of air per minute. It is capable of refilling BA cylinders on site, it carries quite a range of equipment and is also designed to be used as a quick-change area and BA control point.

1975 INTERNATIONAL 1810A
Reg. HO 8014

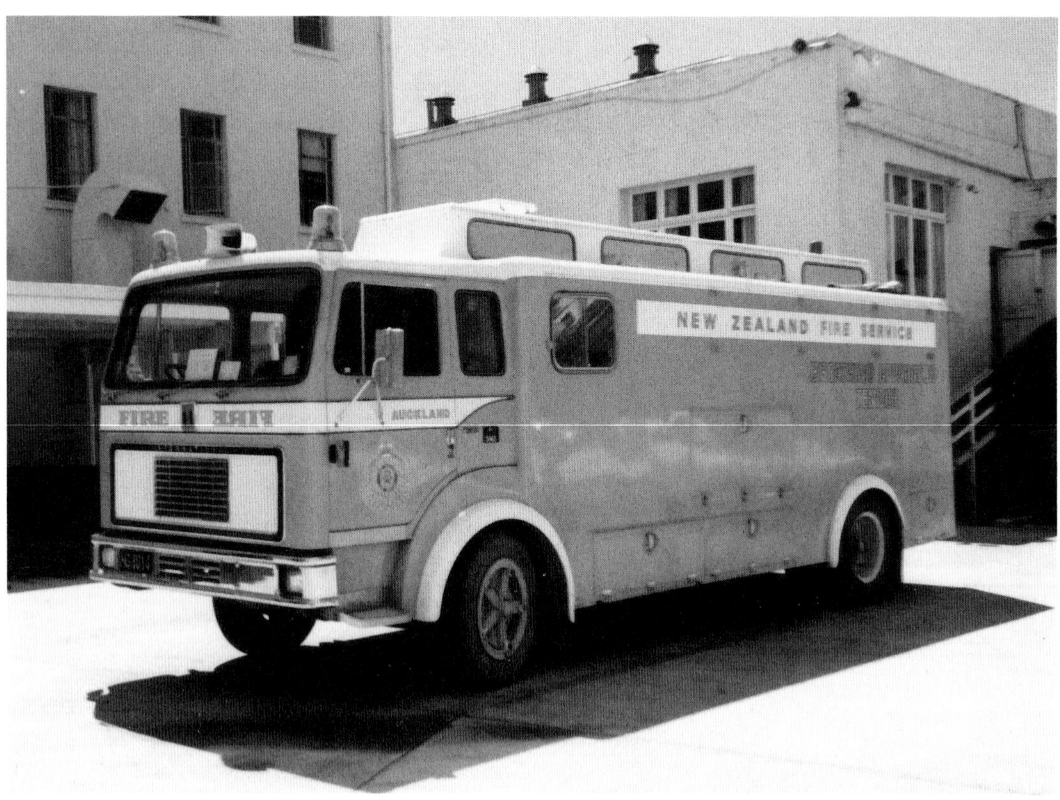

Godiva Pump
Pump cap: 19 L/S
Tank cap: 455 L
Light Rescue
Body mfct: Carmichael

This unusual Range Rover Commando 6x4 is one of three said to have been used by the NZFS. Commissioned in 1975 and then converted to a light rescue unit in 1979, it was stationed in the Wellington area. Disposed of in 1989-90, it was joined by an identical appliance ex-Rotorua which served until 1987-88 when it was also disposed of.

1975 RANGE ROVER
Reg. HW 7478

1978 BEDFORD MK65
Reg. JD 8804

The cab and chassis were brought to Christchurch and the body was built by a local company to the specifications of the local brigade. It carries 66 lengths of 90-mm flaked feeder lines carried on six internal trays, portable pumps, suction hose, portable dam packs, ground monitors, salvage equipment plus various waterway items.

1968 INTERNATIONAL C1500
Reg. DG 1489

Imported by the Auckland Metropolitan Fire Board in 1968, this was one of the early specialised appliances to be used in Auckland, and was in service until 1981. It is now reserve and stationed at Mangere. It carried about 1200 metres of 90-mm hose, two portable pumps, plus an assortment of pumping equipment. It was replaced by a 1980 Dodge (refer to page 134).

This unusual appliance was photographed at Mt Maunganui and is the Tauranga Area Hose-layer. It is affectionately called the 'Bread Box' by the local fire-fighters.

1972 COMMER KCH40
Reg. GE 2571

Commissioned in 1981, this new appliance replaced the ageing 1968 International as Auckland's hoselayer. It carries up to 70 lengths of 90-mm hose flaked in the rear section plus portable pumps and other general waterway equipment. This appliance is capable of conveying water over a large distance to a fire.

1980 DODGE G1513
Reg. JT 620

1981 INTERNATIONAL 1830C
Reg. KE 7625

Waterous Pump
Pump cap: 4 L/S
Tank cap: 3640 L
Tanker
6/21

One of only a few water tankers actually owned by the NZFS, this one is stationed at Palmerston North.

1981 INTERNATIONAL 1830C
Reg. KE 7626

Waterous CPF2 Pump
Tank cap: 3600 L
Water Tanker
6/21

One of two tankers on an International chassis, this is stationed at Fairlie in the Timaru area (the other is at Palmerston North). This appliance has a front-mounted pump and is able to pump and roll as required.

Darley Pump
Pump cap: 30 L/S
Tank cap: 3600 L
Water Tanker
6/21
Body mfct: Mills Tui

Stationed at Russell in the Bay of Islands, this is one of two of this type of appliance built for the NZFS (the other is based at Papakura, on a Dodge chassis). With a front-mounted pump it has provision for fire-fighters to stand on the front platform to 'pump and roll'.

1980 COMMER RG11
Reg. JI 1633

Darley Pump
Pump cap: 26 L/S
Tank cap: 3650 L
Water Tanker
6/21
Body mfct: Mills Tui

A similar appliance but on a Hino cab and chassis. It is stationed at Oneroa on Waiheke Island in the Hauraki Gulf.

1982 HINO FD184K
Reg. KE 3496

1948 LEYLAND
Reg. EZ 1881

Turn Table Ladder 6/10

New Zealand's only 1948 Leyland was a 30-metre Merryweather turn-table ladder placed in service at Parnell, Auckland in 1966 and now at MOTAT. It is said to have been built on a Leyland bus chassis. The ladder arrived in boxes and had to be assembled. One section supposedly had a bend in it as a result of its being left extended in the wind during assembly while the firemen were on their tea break. The photograph was taken in England prior to shipment to New Zealand.

1956 DENNIS F21
Reg. EZ 1882

Dennis Pump
Pump cap: 38 L/S
Turn Table Ladder 6/10

Powered by a Rolls Royce motor, this appliance carried a 38-metre Metz ladder and was in service in Auckland. It is said that on one occasion when it was travelling down a central city incline, the ladder suddenly extended by itself, projecting past the appliance down the hill; apparently the ladder stops failed suddenly. It was preserved in Christchurch with the Fire Service Historical Society but has been placed at MOTAT.

1968 AEC MERRYWEATHER
Reg. DL 3987

Built on an AEC chassis, this appliance has a 30-metre Merryweather turntable ladder. It was still in operation in Dunedin when the photograph was taken.

*Merryweather Pump
Pump cap: 38 L/S
Turn Table Ladder
6/10*

1971 ERF 84PF
Reg. FY 4316

This appliance is stationed at Hamilton and carries the 400th ladder made by Merryweather. In June 1987 the appliance had a new safety cab fitted by Mills Tui. The original cab, being fibreglass, did not meet the required safety standards.

1976 INTERNATIONAL 2150A
Reg. IC 1333

Darley Pump
Pump cap: 57 L/S
Turntable Ladder
6/10

This appliance is fitted with a 30-metre four-section Grove turntable ladder. It was originally supplied with a V8 petrol motor with an Allison auto gear box. It also has a rescue winch and 80 metres of cable with a pull of 3636 kg fitted to the underside of the ladder. This was stationed at Parnell in Auckland but has now been downgraded to the status of a relief ladder.

1978 MACK CF685
Reg. JB 4877

Turntable Ladder
6/10
Body mfct: Rebuilt by Mills Tui

One of the four of its type in use presently, this is based in Christchurch. The others are at Napier and Palmerston North with another on loan to New Plymouth. They carry a 30-metre turntable ladder. All of these appliances underwent refurbishing at Mills Tui in Rotorua during 1992-93.

*Waterous Pump
Pump cap: 130 L/S
Turntable ladder
6/10
Body mfct: Wormald*

Formerly based in Wellington, this appliance is now stationed at Parnell in Auckland. It is powered by a Mack diesel motor and carries a four-section ladder. The ladder, built by Towers Incorporated, can extend to 30 metres and is operated from 6° to 78° inclinations. The appliance is supported by four hydraulic jacks while in operation and can pump up to 133 L/S through the monitor mounted on the top of the ladder.

1978 MACK CF600
Reg. IW 2123

*Metz FP Pump
Pump cap: 44 L/S
Turntable ladder
6/10
Body mfct: Total Walter
Co/Metz*

Especially built for the NZFS by the Metz factory in Karlsruhe, West Germany, the cab and chassis were made by Daimler Benz and the ladders were made by Total Walter Company in Cologne for Metz. There are four of these appliances in New Zealand, with the one shown in Christchurch and the other three in Wellington.

1990 MERCEDES-BENZ
Reg. OM 3920

1975 ERF
Reg. HO 8272

Waterous pump
Pump cap: 55 L/S
Tank cap: 1000 L
plus 250 L of foam.
HEM
6/11
Body mfct: Wormald

The photograph shows the appliance as it was when it went into service in 1976 at Takapuna on Auckland's North Shore. In 1986 it was withdrawn from service due to a chassis fault. It carries a 15-metre Telesquirt and has a 76 litre per second monitor on the end of the telescopic boom. Once set up, the whole operation can be remotely controlled.

1985 HINO FH224
Reg. MH 8861

Waterous pump
Pump cap: 55 L/S
Tank cap: 1000 L
Body mfct: Wormald
HEM
6/11

This appliance is known to have been stationed at Takapuna and Rotorua, and is now at Hamilton. The body, boom and pump from HO 8272–ERF were fitted onto a Hino FH 224 chassis. This has a 6/31 cab and had an Austral plate fixed across the front grille.

Darley Pump
Pump cap: 72 L/S
Tank cap: 1200 L
Teleboom or Telesquirt
(HEM)
6/11
Body mfct: Mills Tui

This is one of three of
this type of appliance
fitted onto an
International cab
and chassis. This one is
based in Whangarei with
the others located at
Christchurch and
Queenstown. The
teleboom/telesquirts are
a pumping appliance
with the special boom
and nozzle for delivering
large quantities of water
on a fire and are thus
ideally suited to high-rise
building fires.

1982 INTERNATIONAL 1950C
Reg. KO 2607

Darley SEH 750 Pump
Pump cap: 55 L/S
Tank cap: 950 L
6/11
Body mfct: Mills Tui
Aerial Monitor

A Scania appliance with
a Sky-Jet hydraulic
elevating monitor, this is
the only one of its type to
date. It is stationed at
Rotorua and had just
arrived from Mills Tui
when this photograph
was taken. The Sky-Jet
HEM was originally on an
International chassis.

1989 SCANIA G93M
Reg. RH 3142

1988 INTERNATIONAL T2670
Reg. NW 6563

Dennis #3 Pump
Pump cap: 60 L/S
PHEP
6/9
Body mfct: Mills Tui

Stationed at Wellington, this would be one of the largest appliances operated by the NZFS. The boom and pump were transferred in about 1988 from its original 1970 Dennis (Reg. FL 3156) cab and chassis by Mills Tui in Rotorua. It is said a few problems were experienced when the brigade first used the appliance due to the number of narrow streets in the Wellington area.

1975 ERF MODEL 84PF
Reg. HU 4939

Waterous Pump
Pump cap: 55 L/S
Tank cap: 750 L
PHEP
6/9
Body mfct: Wormald

This Christchurch-based appliance is powered by a Perkins diesel motor. A similar appliance, also on an ERF chassis, is at Timaru. This one has the polished locker section while that at Timaru has a painted rear section and roller doors on the lockers.

This appliance was stationed at New Plymouth and is understood to be the only one of its kind operated by the NZFS.

1976 INTERNATIONAL 2150A
Reg. IH 8708

1981 MACK CF612
Reg. KC 1509

The first appliance of its kind in the Southern Hemisphere and the largest operating out of the Auckland Central Fire Station, this Mack weighs 22 tonnes and is 12.7 metres long. The maximum boom reach is 19.5 metres. It has two monitors in the basket which can deliver 38 L/S of water on a fire. The basket on the end of the boom can be used at an incident for rescues. The appliance is supported on six jacks while in use and the boom can be operated from either the basket or the control panel on the deck.

1981 MACK MC685
Reg. KL 2414

Darley Pump
Pump cap: 90 L/S
Tank cap: 450 L
PHEP
6/9
Body mfct: Mills Tui

Currently stationed at Manukau, South Auckland, this Mack appliance carries an hydraulic snorkel unit. Capable of reaching 20 metres high it has 360° rotation. A 76 L/S capacity single nozzle is attached to the upper basket. The appliance is stabilised in position with four hydraulic ground jacks. The monitor can be seen in a folded position on the outside of the cage and the ground jacks can be seen in their folded position each end of the rear mudguards. Two first-aid reels are fitted above the pump panels just behind the cab. Both the photographs here illustrate the size of the appliance and the snorkel.

SECTION SIX
Industrial brigades

As well as NZFS brigades, there is also a group of specialist brigades which come under the broad heading of industrial brigades. While they are not under the direct control of the NZFS, they co-operate and assist as required. Like the NZFS brigades, these industrial brigades are members of the United Fire Brigades Association.

Industrial brigades are small in number but have appliances which range from the basic Landrover to the mighty Fire Stryker of the international airports.

Within this group are brigades of the international and domestic airports, the Royal New Zealand Air Force, the New Zealand Army, forestry companies, the Department of Conservation, the New Zealand Oil Refinery, New Zealand Steel, and a number of freezing works and industrial plants.

The industrial brigades have been greatly influenced by the NZFS brigades. As the efficiency and availability of more advanced fire-fighting technology increased, a number of them ceased to function due to the efficiency of the NZFS brigades. Few freezing works still operate their own brigades, and the number of brigades at industrial plants has decreased too. Today the majority of industrial brigades are in more specialised or high fire risk areas. The New Zealand Refinery Company at Marsden Point, for example, has a very modern fleet of appliances for the protection of the refinery plant.

Other examples include the Kinleith Pulp and Paper Mill, the Whakatane Board Mills, and up until recently, General Foods and Alcan Industries.

The major international airports now have their own rescue fire divisions. They all have to provide 24-hour coverage and meet the highest International Civil Aviation Organisation rating of Category 9. While most appliances have only a one- or two-person crew, each has very sophisticated equipment. They must be fast, efficient, reliable and have the capacity to respond immediately to any emergency situation. There have been major changes in these types of appliances over the years. As the aircraft being used by overseas airlines have become larger and flights more frequent, the appliances and equipment have had to develop correspondingly. Whereas once a second-hand Airforce or overseas appliance may have been sufficient, now the appliances are specially designed and built in New Zealand to comply with requirements.

A similar development has taken place in the Royal New Zealand Air Force. They now have a special range of equipment and appliances stationed at each air base. Again, with the larger and more advanced aircraft used, they have developed their own special appliances.

While this development has not been as fast in forestry and the Department of Conservation, the use of the helicopter and monsoon buckets has changed fire-fighting techniques. Heavy duty four-wheel-drive vehicles are relied on to transport the basic equipment and many of the early model appliances are still being used today.

The New Zealand Army has at its main military camps its own trained fire personnel and equipment. While they mainly service their own buildings and areas, they are also available to assist NZFS brigades when required.

1953 FORD MARMON
Reg. DV 3281

Colmoco Pump
Pump cap: 33 L/S
5 L/S of foam
Tank cap: 2270 L
Airport Crash Tender

Several of these appliances were built for the RNZAF. This one spent most of its life at the Wigram Base. Powered by a V8 motor it carries a large tank for protein foam which is mixed by a proportioner at the pump. It was given to the Fire Service Historical Society in 1971 and has since been restored to its present condition.

1954 R BEDFORD
Reg. EZ 4534

Gwynne/Colmoco Pump
Pump cap: 38 L/S
Tank cap: 4550 L
4x4 Tanker

Used by the RNZAF as a tanker to back up the crash tenders. Some 227 litres of foam were carried in a rear tank and a small amount of equipment in the lockers. A slow machine, about five or six were in service until 1987. This was once stationed at Whenuapai.

Used by the RNZAF as a normal structural appliance, this carried the usual waterway equipment, hose and breathing apparatus units. It also carried four drums of foam. It was originally numbered SV 177 before being re-numbered 18. It was stationed at Woodbourne and remained in service until 1987. Similar appliances were at Whenuapai, Ohakea and Wigram.

1957 S BEDFORD
Reg. DY 2677

Twelve of these RNZAF aircraft crash fire vehicles were stationed at the various bases; this one was at Woodbourne. They had a 568-litre foam tank mounted over the rear mounted pump and had a Pyrene 20x roof monitor, controlled by a fireman standing through the roof hatch. They were fitted with crash bars and a hook for driving through gates or fences. All controls were manual.

1964 INTERNATIONAL AA160
Reg. EV 4394

1976 BEDFORD J6

Reg. HW 8890

Coventry Climax
Pump cap: 57 L/S
Tank cap: 900 L
Training Appliance
Body mfct: Wormalds

Used by the RNZAF as a training appliance at No. 3 Technical Training Fire School, this was an ex-New Zealand Army appliance which had served at Waiouru and Linton Military Camps. It was decommissioned in 1994. Photographed at Wigram Base.

1983 DODGE POWER RAM

Reg. LE 2811

These four-wheel-drive appliances were used by the RNZAF at most bases. Powered by a 360 V8 with auto-transmission, they were used as RIU (Rapid Intervention Units). This unit was based at Whenuapai near Auckland.

The cab and chassis were manufactured by Unipower Vehicles Ltd of England with the body being built by Mills Tui in Rotorua. It is powered by a Detroit turbo-charged V8 12-litre two-stoke diesel motor developing 410 kW and has an Allison HT 740 four-speed auto gear-box. It can accelerate from 0 to 88 km/h in 22 seconds. Maximum speed is 120 km/h. Appliance 22 is based at Whenuapai.

UNIPOWER 'FIRE STRYKER'
Reg. MK 8083

This has a Feecon roof monitor and can be hydraulically or manually controlled from the cab. It has a throw of 70 metres at 1400 kPa. One of these vehicles is located at each RNZAF base and kitted out as a residential appliance for normal fire duties. Appliance 04 is seen here at Ohakea.

UNIPOWER 'FIRE STRYKER'
Reg. NJ 4339

1983 INTERNATIONAL 1810

Reg. KT 4853

Darley SEH 500 Pump
Body mfct: Mills Tui

Stationed at the Waiouru Military Camp, this appliance is kitted out as second response to 'in camp property' calls and carries normal domestic fire fighting equipment.

1983 INTERNATIONAL 1810

Reg. KT 4852

Darley SEH 500 Pump
Pump Rescue
Body mfct: Mills Tui

Based likewise at Waiouru Military Camp, this appliance is used for 'out of camp' domestic calls. It carries Viking chemical suits for chemical incidents, a Riche rescue kit and Roberts cutting gear, plus an Angus portable pump.

Eighteen of these appliances were built for the Ministry of Defence and issued to the NZ Army and the RNZAF. (The Ministry of Defence is the Fire Authority for all defence lands gazetted as a Rural Fire Districts.) It is built on a four-wheel-drive Hino chassis and has special facilities for aerial refilling of the water tank. Forest foam production and a front facing monitor for running fires are controlled inside the cab. It is designed for a two-man crew and has an overnight sleeping bunk for one person.

Heavy Pump
Body mfct: Mills Tui

Built to the same design as the NZFS 6/3 Heavy Pump. Used by the NZ Army in most of their military camps, this one is based at Linton Military Camp just south of Palmerston North.

1989 4x4 HINO GT175
Reg. NO 174

1991 SCANIA G93M
Reg. PK 8394

1942 FORD V8

Reg. DV 5144

This was originally an army scout car before being purchased by the Forestry Service and converted. It carries no water but has a transverse mounted pump unit and carries suction and a coiled hose. It was stationed at Eyrewell Forest Headquarters before being purchased by the Fire Service Historical Society in 1990.

1942 FORD V8

One of about 27 such appliances once scattered throughout New Zealand. Converted from an army scout car by the Forestry Service in about 1946, it is powered by a side-valve Ford V8 motor positioned above the rear axle. It has a dual-range gear box, two- or four-wheel-drive transfer box and a four-speed main box.

A further example of the forestry four-wheel-drive water tanker. This one is fitted with a first-aid hose reel, a larger screen and the luxury of windscreen wipers. It was in service in the Canterbury Forest.

1942 FORD V8
Reg. CX 4674

This version of the four-wheel-drive Bedford was typical of standard forestry appliances of the late 1970s and 1980s. It usually carried an assortment of hose including the 'forestry packs' which were normally two lengths of hose, flaked and fitted into a back-pack. These are still carried by a number of appliances today. This one was photographed at Naseby Forest in the South Island.

1978 BEDFORD TK
Reg. IT 5016

1983 ISUZU
Reg. LH 3902

Operated by New Zealand Forest Products in Northland, one each of these four-wheel-drive appliances is based at Dargaville, Warkworth and this one at Kauri, just north of Whangarei.

1985 ISUZU JCF
Reg. MA 2146

Built to the specifications of the Department of Conservation by Mills Tui on a 4x4 Isuzu chassis, a number of these operate throughout New Zealand in areas under the Department's control. A monitor is mounted on the rear. This one, based in Kaikohe in Northland, additionally carries a monsoon bucket.

These appliances were used extensively around major forest areas. This one is operated by New Zealand Forest Products, Kinleith. They had a rear crew compartment and normally used a portable pump fixed on the rear deck which enabled them to pump and roll at the same time.

1970 BEDFORD TK
Reg. IY 3174

Several of these tankers are used by the larger forestry companies. They often use the tractor units from the logging trucks. These examples were photographed at New Zealand Forest Products at Kinleith.

1983 KENWORTH
Reg. NJ 4483

1953 MARMON HARRINGTON

Reg. DV 226

A number of appliances were built to this design, supplied by the Department of Aviation and used at domestic airports around New Zealand during the 1950s. This one was in service at Christchurch airport as MOT No. 212. It has been repainted its original white colour and is under the care of the Fire Service Historical Society.

INTERNATIONAL AA160

Reg. EU 4197

Airport Crash Tender

Another version of the early airport crash tender is this one based on an International chassis. This was once in service at the Auckland Airport as MOT No. 477. It is now at the North Shore air club at Dairy Flat and in private ownership.

Two of these appliances were based at Rotorua Airport, operated by the Airways Corporation of New Zealand. Both had fibreglass cabs. These have since been replaced and are no longer in service.

1980 HOYLE ARRESTOR
Reg. JP 5857

An appliance built on a Scania chassis with a fibreglass Bedford-style cab, this was based at Whakatane Airport as old MOT No. 499. It was rebuilt by Mills Tui into a training appliance for Christchurch Airport (refer to page 164).

TRUCTOR ARRESTOR
Reg. JE 5700

1972 INTERNATIONAL D400
Reg. GB 1686

Built on a 6x6 International chassis by Wormalds, this appliance was stationed at Auckland Airport. A similar appliance, registered as GB 1687 (MOT 410), was stationed at Christchurch.

1979 INTERNATIONAL
Reg. JE 5698

An early model crash tender which was based at Wellington Airport when it was under the auspices of the Ministry of Transport.

Waterous Pump
Pump cap: 79 L/S
Tank cap: 11,250 L
990 L AFFF (foam)
Crash Tender

This Mack was powered by a turbocharged V8 diesel engine through a 10-speed split-ratio gear box. A separate 280 kW Caterpillar V8 motor powered the waterous pump which gave the appliance its pump and roll capacity. Six of these 6x6 models were intended for use at major airports. The photograph was taken at Auckland Airport.

1974 MACK FR700
Reg. HF 2015

Waterous Pump
Pump cap: 79 L/S
Tank cap: 11,250 L
990 L AFFF (foam)
Airport Crash Tender

A sister to the previous appliance, this was based at Christchurch Airport and is now at Rotorua Airport. All of these types of appliance carried the new type of foam called Aqueous Film Forming Foam (AFFF). This is a synthetic foaming liquid which can extinguish a flammable liquid fire in half the time of conventional foam. It spreads over the surface of burning fuel, forming a blanket-like conventional foam. An aqueous solution, however, drains from the foam bubbles and forms a vapour-sealing film.

1974 MACK FR700
Reg. HW 4932

WALTER
Reg. JO 6806

Tank cap: 1135 L
plus AFFF (foam)
Dual Purpose Vehicle

Built by Wormalds, this appliance was a dual purpose vehicle for the RNZAF and stationed at Ohakea Airforce Base as No. 25. It is now based at Palmerston North Airport for United Aviation Services.

4x4 HINO
Reg. ND 8134

Waterous Pump
Tank cap: 3000 L
500 L AFFF foam
Body mfct: G.A. Zander,
Palmerston North

Owned by United Aviation Services, this crash/rescue tender is based at Palmerston North Airport. It has a Rosenbauer roof-mounted monitor and is capable of one-man operation from the driver's seat. A twin-cylinder BCF system with hose and reel is mounted in the rear locker. They have retained the traditional white and yellow colour scheme.

1988 HINO
Reg. MT 7740

This appliance is in operation at Wanganui Airport. It is built on a Hino chassis and is a four-wheel-drive vehicle. Fitted with both water and foam, first aid reels and a roof monitor, it has been designed for one-man operation.

TRIDENT AUSTRAL
Reg. OK 7877

One of the three Trident appliances built by Austral in Australia for the new Austral-Armourguard joint venture at Wellington Airport. Two of them are the MK5 model, powered by a 403-kW Detroit engine. They carry a minimum of 7000 litres of water plus AFFF compound. They are capable of total operation by one person from the driver's seat. The third appliance is an MK6 and designed as a Rapid Intervention Vehicle (RIV). This carries only 5400 litres of water, specialised equipment, BCF and dry power systems, specialist equipment plus a larger crew.

ROSENBAUER JUMBO CHEETAH

Reg. MR 8663

Rosenbauer R280 Pump
Pump cap: 1600 L
11,200 L foam
Tank cap: 200 L
20 L foam
Airport Rescue Tender

Stationed at Auckland Airport, this appliance is powered by a MAN 440 H turbocharged V10 diesel engine. It has 4x4 auto-transmission, with a roof monitor. Rescue One carries a crew of one officer and two rescue fire fighters. It has first aid equipment, resuscitator, hydraulic rescue gear, pneumatic cutting gear, BA, portable generator and flood light. All appliances are fitted with a variety of radios for communications.

1989 STRYKER 6

Reg. OS 6827

Darley PSE 1500
Pump cap: 5000 L
50,000 L foam
Tank cap: 9100 L
1200 L foam
Air Rescue Tender
Body mfct: Mills Tui

This is one of four appliances of this type operated by the Auckland Airport Rescue Fire, powered by a Detroit 540 H turbocharged V8 diesel engine with 4x4 and 6x6 auto-transmission. It has a separate Detroit engine to power the pump and the roof and bumper monitors. It also carries BA units, rescue tools, standpipe, hose and fire fighting branches, rescue saw and portable extinguishers. The monitors can be operated from a console in the driver's compartment.

Built by Mills Tui circa 1990 as one of a batch of new training appliances for Christchurch Airport. The chassis is said to have come from the Truktor Arrestor (Reg. JE 5700) previously based at Whakatane Airport, MOT No. 499. (Refer page 158).

MACK/SCANIA
Reg. RE 6412

1992/93 STRYKER 6/2
Reg. RE 6695

This vehicle has a Spartan chassis with six-wheel-drive and twin rear-mounted Detroit V8 turbocharged diesel engines, a top speed of 120 km/h, and can reach 0-80 km/h in 35 seconds. The second motor powers the pump. A Feecon non-aspirated short-barrelled monitor is turret-mounted on the roof and has an output of 79 L/S to a distance of 78 metres. A second monitor is bumper-mounted and both are controlled from the cab. Both monitors plus the three below-vehicle nozzles can work at optimum pressures simultaneously.

1981 INTERNATIONAL 2150B
Reg. FIRE 1

Darley Champion
N1500 Pump
Pump cap: 114 L/S
Tank cap: 4500 L foam
Aerial Monitor
Body mfct: Mills Tui

Originally built with a crew cab and deck monitors, this appliance was rebuilt to its present form by Mills Tui in 1987. It carries a 15-metre Aerial Monitor (Snozzle) capable of delivering 75 L/S from the nozzle. It also has eight deliveries for foam and/or water. The monitor can be remote-controlled from the rear of the appliance.

1984 NISSAN CW20
Reg. FIRE 2

Waterous Pump
Pump cap: 120 L/S
Tank cap: 4500 L
of foam
Foam/Water Pump

An important appliance for any major petroleum incident, with eight deliveries, this appliance is fitted with a Total foam/water monitor which is capable of throwing jets of either up to 60 metres at 1000 kPa. Both of these appliances are based at the Marsden Point oil refinery.

This appliance is based at the New Zealand Refinery complex at Marsden Point near Whangarei. Its foam capacity and Total monitor capable of 1000 kPa are essential in the control of petroleum fires.

1985 NISSAN CW20
Reg. FIRE 3

This is a fire control unit for all major incidents at the Marsden Point refinery. It carries all maps and plans of the process area and tank farm. It has been fitted with complete radio systems for both the New Zealand Fire Service and all refinery channels.

BEDFORD J1
Reg. DZ 9523

TOYOTA LANDCRUISER
Reg. LL 8445

This vehicle was used for emergency situations and security at Marsden Point and has now been replaced. It carried a range of equipment similar to that of an emergency tender. This photograph was taken while it was deployed at Marsden Point.

NISSAN PATROL
Reg. MB 7183

Emergency Tender

Also stationed at the Marsden Point oil refinery, this tender carried a full range of rescue and other equipment plus spare breathing apparatus sets and was also used for security patrols. It has since been replaced by a Mitsubishi L300.

Coventry Climax
Pump cap: 38 L/S
Tank cap: nil
Body mfct: Mills Tui

This unit is used in major
incidents at the Marsden
Point oil refinery to pump
water or foam.

1982 TRAILER PUMP

Armag Pump
Light Pump

1976 LANDROVER

Reg. EW 6044

Stationed at the Howick
Volunteer Fire Brigade in
1976, this vehicle was
later transferred to the
Papakura Volunteer Fire
Brigade. It was finally sold
to Alcan Industries in Wiri,
South Auckland, for their
industrial brigade and
disposed of when their
brigade was disbanded.

1955 DENNIS F8

Reg. EZ 1890

Dennis Pump

One of several of the type imported by the Auckland Metropolitan Fire Board during the 1950s. It was used from 1955 to 1974 at a number of Auckland brigades. Following this, it was transferred to Kawakawa Bay in South Auckland. Finally, in 1981, it was sold to New Zealand Railways for their Otahuhu workshops (refer to page 184).

1955 DENNIS F12

Reg. DZ 6395

Dennis Pump
Pump cap: 62 L/S

This appliance is thought to have been originally stationed in Auckland or on the North Shore. It was transferred to Whangarei and is known to have been stationed at Ruakaka. It was finally deployed at the AFFCO Freezing Works at Moerewa and was still in service until late 1993 when it was sold. It once carried a wheeled escape.

Acquired by the Auckland Metropolitan Fire Board in 1962 and stationed at Headquarters and then as No. 1 at Mt Eden and back-up to Headquarters. In 1970 it was moved to Mt Wellington. It was sold to AFFCO for their freezing works at Rangiuru in 1982 and is used it as a stationary pump, permanently 'plumbed in' to the water main.

1962 KARRIER CARMICHAEL
Reg. EZ 1876

Originally owned by the Napier Fire Board. It was sold to the Whakatane Board Mills Industrial Brigade in Whakatane, operated by New Zealand Forest Products Pulp and Paper Ltd. The body has been altered a number of times but it is still in mint condition after more than 40 years of operation.

1951 FORD V8
Reg. EN 6046

1988 MACK MAINLINER
Reg. NS 7312

Darley SH 1250
Pump cap: 80 L/S
Tank cap: 1300 L
169 L of foam
Heavy Pump
Body mfct: Mills Tui

Operated by the Kinleith Industrial Brigade at the pulp and paper mills there, this is powered by a six-cylinder Turbo diesel through an Allison automatic transmission. The rear tank and locker module was made to New Zealand Fire Service 6/3 specifications. It has a Santa Rosa monitor mounted on the locker section roof. An impressive machine, it stands out with its unusual colour scheme and roof-mounted light bar.

1971 INTERNATIONAL C1800
Reg. FQ 9606

Darley Pump
Pump
Body mfct: Mills Tui

Back-up pump to the Mack operated by the Kinleith Industrial Brigade, this machine is also painted in the same colours, i.e. white with red stripe and details. It is said that, due to the chemicals in the area, the usual red paint scheme required greater upkeep than the white does.

Portable Pump
Body mfct: Local

Stationed at the Taharoa Iron Sands complex, a division of New Zealand Steel Mining Ltd, at Taharoa on the west coast of the North Island just below Raglan, this appliance carries a variety of equipment including a portable pump, BA units, hose, etc. It is a support vehicle for their other main pump.

1985 TOYOTA LAND CRUISER
Reg. LK 5329

Darley Pump
Body mfct: Mills Tui

This is the main appliance for the Taharoa Industrial Brigade at the iron sands plant. The four-wheel-drive vehicle is ideally suited to the rugged area. Its cab and chassis were originally part of a top-dressing truck.

1976 MERCEDES LA1113
Reg. HN 1726

1960 BEDFORD

Reg. IR 4993

An ex-Ministry of Transport airport tender, this is now based at the Huntly Power Station Industrial Brigade. It is used mainly around the complex and is backed up by the Huntly Volunteer Fire Brigade in any major incident.

MACK

Reg. LW 8959

Two of these appliances are operated by Synfuel as part of their Industrial Brigade. The other appliance, also a Mack, is slightly smaller and has only one rear axle. Both have rear deck monitors. No details are available as to the pump or tank capacities.

Fire parties and appliance restoration

Fire parties have been an important part of overall fire management for many years. A fire party is a smaller brigade which typically operates in an area of low population density and comes under the jurisdiction of the local district council rather than the New Zealand Fire Service. A number of brigades operating today under the New Zealand Fire Service started as fire parties. As a result of organisational changes new fire parties are starting up, and existing fire parties are obtaining better equipment.

Recent changes in legislation have seen a change in the role played by the New Zealand Fire Service Commission. It now has two distinct functions. It controls the New Zealand Fire Service, and it is also the National Rural Fire Authority. The National Rural Fire Authority co-ordinates rural fire control and codes of practice, promotes training and makes grants to Fire Authorities.

The lands generally protected by Fire Authorities are National Parks, crown land and conservation areas administered by the Department of Conservation. Rural Fire Areas are areas where fire control is the responsibility of a rural fire committee. An area which is neither covered by any of the above nor by an existing New Zealand Fire Service brigade is the responsibility of a district council or city council.

With the replacement of New Zealand Fire Service appliances, many of its older appliances have been sold off to a number of fire parties. This has meant that the fire parties are able to obtain equipment that otherwise they would never be able to afford. A number of district councils have obtained water tankers which are either manned by the local volunteer brigade or by their own staff.

In New Zealand there are a number of enthusiasts who have managed to obtain vintage appliances. Some volunteer brigades have been able to obtain their original appliances and, after many hundreds of hours of work, restore them to their former glory.

1971 INTERNATIONAL C1600
Reg. FT 4331

Gwynne Pump
Light Pump

This is an ex-NZFS appliance, previously stationed at Piha and Waiatarua in the Waitakere Ranges. It was nicknamed 'Dumbo' and was powered by a six-cylinder engine, replaced later by a V8 motor. The vehicle is now in service with the Paparoa Fire Party and operates from their station at the Otamatea District Council yard.

1965 BEDFORD TK
Reg. DN 6558

Coventry Climax Pump
Pump cap: 38 L/S
Tank cap: 680 L
Light Pump
Body mfct: Mills Tui

Repowered in 1978 with a Holden V8 engine, this appliance was once stationed at Papatoetoe Volunteer Fire Brigade and then at Russell Volunteer Fire Brigade. It is now at Towai (between Whangarei and Kawakawa) and is under the control of the Towai and Districts Fire Party.

Said to have been based originally at the Morrinsville Volunteer Fire Brigade, this appliance was found behind a store at Hatepe, just south of Taupo. It is now owned by the Taupo District Council and operated by the Hatepe Fire Party.

BEDFORD J3

Reg. EH 2660

Coventry Climax Pump
Light Pump
Body mfct: Prestige Ltd,
England

This was originally owned by the Waitemata County Council and stationed at Huia in the Waitakere Ranges west of Auckland. It is now with the Mangamuka Fire Force as their pump. It is used to fight scrub fires in a large rural area as well as to deal with motor vehicle accidents on State Highway 1 just south of Kaitaia.

1957 BEDFORD A3

Reg. EH 1531

1965 BEDFORD J2
Reg. EQ 9609

Portable Pump
Rural Fire Appliance

Built on a Bedford J2 standard chassis, this carries the basic equipment for fighting rural fires. It is operated by the Koitiata Fire Party at Turakina Beach just south of Wanganui, which comes under the authority of the Wanganui District Council.

1958 FORD V8
Reg. DY 8894

Colmoco Pump
Body mfct: Colmoco

This appliance was stationed with the Houhora Volunteer Fire Party in the Far North and was in service there until 1993 when it was sold in mint condition to a private collector. It is said to have served in the Kaitaia area previously.

1955 BEDFORD S
Reg. EJ 2275

Belonging to the Far North District Council, this appliance is used by a new fire party at Broadwood. Previously it was used by the Rangiputa Fire Party on the east coast just north of Kaitaia, as well as at Houhora. It was originally a Crash Fire Tender and stationed at the RNZAF Base at Te Rapa.

1966 BEDFORD
Reg. FR 1766

An ex-New Zealand Army appliance, once stationed at the Linton Military Camp, this machine is now used by the Bethells Valley Fire Force on the west coast of Auckland. It plays an important role in the area due to the shortage of water, especially in the summer season. The area, with its beach, is a popular holiday resort.

1954 LANDROVER SERIES 1
Reg. NW 8787

Portable Pump
Light Pump

This appliance, called 'the smoke chaser', is operated by the Lake Tarawera Fire Party in the Rotorua area. It carries a portable pump and a small amount of other equipment. It is one of two appliances used by this Fire Party, the other being a Karrier Carmichael.

1958 LANDROVER SERIES 1
Reg. DY 8893

Light Pump

This unusual yellow-painted Landrover is part of the Paparoa Fire Party fleet and is located between Ruawai and Maungaturoto on State Highway 12. It is an ex-NZFS vehicle, acquired from the Waipu Volunteer Fire Brigade.

Operated most recently by the Karekare Fire Party on Auckland's west coast, this unusual appliance was originally based in Lower Hutt and is said to have been a St John's rescue unit. It was also used by the Lower Hutt Brigade and as a control unit by the Auckland Brigade before being transferred to the Fire Party. It has since been replaced by a new rural fire appliance.

1965 LANDROVER

Reg. FB 4304

Rural Fire Appliance

This appliance was built by members of the Mangonui Volunteer Fire Brigade. It is now stationed at Houhora Fire Party at Pukenui in the Far North, one of a number that comes under the control of the Far North District Council.

1975 BEDFORD TK

Reg. HF 2966

1981 JEEP J20

Reg. KN7114

Robin Portable Pump
Tank cap: 1350 L
Rural Fire Pump
Body mfct: Mills Tui

Purchased for the Muriwai Volunteer Fire Service, which is now controlled by the Rodney District Council. The rear lockers carry the basic equipment and it has a portable pump permanently plumbed in on the rear. A first aid reel is located behind the cab. The four-wheel-drive is ideally suited to the Muriwai Beach area.

1983 BEDFORD TK

Reg. KR 5283

Portable Pump
ex-Forestry Appliance
4x4

This appliance is used by the Whangaruru Fire Party which is located at Oakura on the east coast of the North Island between Whangarei and Russell. It is thought to have been acquired from the Russell forest area.

*Darley 2½ Age
Portable Pump
Tank cap: 1000 L
Rural Appliance
Body mfct: Mills Tui*

One of the first specially designed rural fire appliances — by Kevin O'Sullivan for the Waitakere Council — this is said to have cost about half as much as a 6/1 NZFS Light Pump and has accordingly been dubbed 'six-half'. The emphasis was on simplicity, using a crew cab Daihatsu with a specially designed locker and water tank. It has a pre-connected portable pump with two 30-metre lengths of Duraline connected and stored flaked in a top locker. A second unit is now also stationed at Karekare.

*Dennis Pump
Pump cap: 30 L/S
Tank cap: 680 L
Light Pump*

Three of these machines were purchased by the Far North District Council from the NZFS; this one, now sporting a yellow paint scheme, came from the Russell Volunteer Fire Brigade. It is now based in Kaitaia and used by the local council as part of their fire-fighting facilities.

1993 DAIHATSU DELTA
Reg. RQ 8759

1974 TOYOTA LANDCRUISER
Reg. GH 3356

1958 KARRIER CARMICHAEL

Reg. EZ 1869

This appliance started service under the Auckland Metropolitan Fire Board in 1958 at Parnell. In 1965 it was transferred to Onehunga, then in 1970 to Headquarters Relief. It was finally sold to the Huntly Volunteer Fire Brigade. After withdrawal and a few subsequent private owners, it was found by its present owners in a shed north of Kaikohe.

1958 KARRIER CARMICHAEL

Reg. BASH 66

This is the same appliance as above after restoration by a group in Whangarei as a promotional vehicle. The old Super Snipe motor was replaced by an Isuzu diesel motor and coupled to an Allenson Automatic Transmission. It was used in the 1994 South Island 'Variety Bash' and is called 'The Spirit of Northland'.

Originally based at the New Zealand Railways workshops in Otahuhu, South Auckland (refer to page 169), this appliance was later restored and is now one of the many in the annual 'Variety Bash'. It has been painted in yellow and carries the logos of its sponsor.

1955 DENNIS F8

Reg. AMI 1

Another example of an appliance now used for promotional work and the 'Variety Bash', this vehicle was originally stationed at Ohope in the Bay of Plenty.

1964 BEDFORD TK

Reg. EL 456

1935 DENNIS ACE

Reg. EP 3197

This appliance was purchased by the New Plymouth Fire Board in 1935 and commissioned in 1936. The body was built of kauri wood in Christchurch. The appliance does not carry a hose reel or any water, but has reciprocating primers and small diameter suction hose. After serving at New Plymouth for 30 years it was purchased by Roebuck Construction in 1965, who sold it to New Plymouth Fire Station Officer Willie Wood in 1972 with only 5280 miles (8500 km) on the odometer. Willie Wood first met the engine he was to own and restore in the early 1940s when it came to his school. He could not resist the lure of the engine, got a little too close, and received a curt word from his teacher. However, the lure of the engine was just as strong 30 years later and he purchased his 'pride and joy'.

Restoration of this appliance took a lot of time and patience, and Mr Wood also had to complete the work in time for parades in Wanganui, Stratford and New Plymouth. Extensive chrome work was required, along with re-upholstery of seats, removal of panel dents and a new paint job complete with a gold-leaf border and lettering in the original style. Thanks to the help of another worker who assisted when he had no other duties around the fire station, the engine was

(1935 Dennis Ace
continued from previous
page)

completed on time.
Some trouble was
encountered with the
clutch. It turned out that
modern oil is of too high
a quality for the vintage
vehicle, and when oil of
less purity was used, the
clutch worked without
difficulty. The engine
took 'pride of place' in
the New Plymouth
Jubilee Parade as well as
in the other parades.

1938 FORD V8

Reg. FLICK 1

Two years involving countless hundreds of hours went into the restoration of this appliance, which is identical to the first one in Cambridge. The actual first appliance was located by the group owning this one, but they could not afford to purchase or swap. A new wooden body was built and the cab was stripped and repainted, but it still has the original tyres. Most restoration work was done by brigade members with help from local businesses. It is now the pride and joy of the Cambridge Volunteer Fire Brigade; an asset to the town and a piece of history that could have been lost forever.

1946 FORD V8

Reg. ID 6387

This Ford was placed into service in Taumarunui in 1949 and served there for almost 18 years before being transferred to Owhango (20 km south of Taumarunui). It was sold at auction in Rotorua in 1979, to disappear. In 1987, Taumarunui brigade members began the long task of its location. It was finally found in Gisborne and while the body was rather bent and buckled, the motor was still good and it still had its original tyres. It was driven back to Taumarunui and with a bit of help from the local community and business, was restored to its former glory. It is now used for promotional purposes.

1943 FARGO

Reg. DZ 1986

Found hidden behind a garage in Ruawai, this originally began service for the Auckland Metropolitan Fire Board in 1943 at Ellerslie. In 1957 it was sold to Ray Vincent Ltd. It eventually went back into service with the Ruawai Volunteer Fire Brigade and then the Houhora Volunteer Fire Brigade, before being brought back to Ruawai. It is in the process of being restored to its former glory.

1936 FORD V8

Reg. 36 FIRE

Members of the Morrinsville Volunteer Fire Brigade have restored this, their original 1936 Ford appliance. It is used for parades and special occasions where it takes pride of place.

1934 FORD V8

Reg. FE 1450

First brought into service by the Matamata Volunteer Fire Brigade in 1936 and purchased from them in 1964, this was the first appliance to be used by the Ngatea Volunteer Fire Brigade. It was restored in 1992 by brigade members. It has a 1936 V8 trailer pump.

1966 INTERNATIONAL

Reg. HQ 6884

Tanker

It is believed this may have been brought to New Zealand by the US Airforce as a crash tender. It was bought by the New Zealand Fire Service from the MOT who used it at Auckland Airport. It was subsequently located in Taupo after being used to help with a flood in Te Aroha on 17 February 1985 and was later purchased by the Gisborne District Council. On the way to a call-out it unfortunately spun out on a corner and rolled. No one was injured, but it was badly crushed. It is not known what happened to it after this photograph was taken in the Gisborne District Council Depot.

Stationed with the Rolleston Volunteer Fire Brigade, this tanker was built by the brigade members on a cab and chassis transferred from Southbridge Brigade in early 1989. The tank was donated by Caltex and has an Angus pump plumbed into the suction inlet. It has two domestic and two forestry outlets on each side, a front forestry outlet and spraybar plus a first aid reel. A hydroblender is incorporated into the forestry circuit. All equipment was supplied by the brigade except the pump.

1968 FORD D800
Reg. DL 7408

This tanker was built by members of the Dannevirke Volunteer Fire Brigade with the support of local business people. The tank was donated by Shell Oil Company. Powered by a 270 hp (362 kW) Turbo Isuzu motor with a 13-speed road ranger gear box, it is owned by the brigade and all running costs are covered by the sale of water to farms etc.

1981 ISUZU
Reg. OK 1248

1972/74 BEDFORD

Reg. RD 9174

Petrol Portable Pump
Water Tanker

Another tanker built and used by the Mangawhai Heads Fire Party. The tank was fitted on a cab and chassis supplied by the local district council. It carries a portable pump and has a first aid reel fitted to the rear.

1965 INTERNATIONAL AA160

Reg. EZ 4031

Portable Pump
Tank cap: 2050 L
4x4 Rural Appliance
Body mfct: Forestry

A four-wheel-drive ex-forestry tanker, now used by the Muriwai Volunteer Fire Force.

1972 BEDFORD TK

Reg. FW 7388

Owned by the Selwyn
District Council, this was
built by the members of
the Southbridge Volunteer
Fire Brigade. The cab
was formed by welding
two cabs back to back. It
has a boom fitted across
the front and can be
operated by the driver,
ideal for crop and stubble
fires. It also has special
short deliveries which are
used to fight fires by
standing on the back of
the tanker. The lack of
doors in the rear of the
cab is said to make
conditions a little draughty
for crew.

1965 BEDFORD

Reg. EQ 8400

The cab and chassis for
this tanker came from the
Rangitikei Power Board.
The tank and locker units,
which were built by the
members of the Lincoln
Volunteer Fire Brigade in
1964, were taken from
their original 1950 OB
Bedford appliance and
refitted to this unit.
Owned by the Ellesmere
County Council, and later
the Selwyn District
Council, this served from
1981 to 1991 when it
was sold to the Rakaia
Volunteer Fire Brigade.

1972 BEDFORD MK5

Reg. GB 7971

Browns Bros Pump
Pump cap: 15 L/S
Tank cap: 4500 L
Water Tanker

An appliance stationed at Lake Tekapo Volunteer Fire Brigade, this is now owned by the McKenzie District Council. Previously it was owned by the Forestry Service and based at Nelson.

1966 FORD D750

Reg. DW 3162

Centralised Pump
off PTO to Low
Pressure Hose Reel

Purchased in 1979 by the Portobello Volunteer Fire Brigade, who still man it, although the running expenses have been assumed by the Dunedin City Council Rural Fire Authority since July 1993. It has been fitted out with a tank (from the Mobil Oil Company), lockers and an Angus 1200 portable pump permanently connected to the tank. It saw its first call-out in December 1979.

Pump cap: 8 L/S
Tank cap: 8200 L
Tanker

1967 COMMER TS3

Reg. EP 6660

This vehicle, previously a milk tanker, is owned by the Stratford Volunteer Fire Brigade who receive financial assistance from the District Council for its maintenance. The tractor unit has since been replaced by a 1983 Isuzu JCR due to a major motor failure.

1982 ISUZU

Reg. KO 950

Owned and operated by the Ohakune Volunteer Fire Brigade, this ex-Dairy Company tanker is used regularly, especially during the summer period.

Acknowledgements

I wish to express my gratitude and appreciation for all the help and assistance given during the research for this book.

To my fellow fire appliance enthusiasts for supplying a number of photographs: Tony Frost (Hamilton), Peter Ashley (Hastings), Tom Holder (Christchurch), Nigel Capon (Auckland), Peter Rietveld (Dunedin), Graeme Campbell (Christchurch), Warren Fraser (Hastings), Gary Walker (Auckland) and Keven Farley (RNZAF).

A special thanks to Nigel Hope of the Fire Services Historical Society for his assistance and photographs of the society's collection of appliances.

To the staff at MOTAT, particularly John Walker.

To the Alexander Turnbull Library for allowing the reproduction of some of their photographs.

To Catherine Chapman and Ngaire Wingate of New Zealand Fire Service National Headquarters in Wellington for their assistance.

To the many volunteer and permanent fire personnel who assisted by opening their stations and trundling out their appliances for photographing. To all those brigades that assisted by supplying information on past and present appliances.

To my faithful 'scribe' who chatted up firemen for information and details. To the typist for her assistance and patience throughout the project.

And finally to Geoff Churchman of IPL Publishing for making it all happen.

ABBREVIATIONS

BA	Breathing Apparatus
cap	Capacity
Colmoco	Colonial Motor Company
EFS	Emergency Fire Service
ET	Emergency Tender
Haz-mat	Hazardous material
HEM/PHEM	Hydraulic Elevating Monitor
HEP/PHEP	Hydraulic Elevating Platform
km/h	Kilometres per hour
L/S	Litres per second
MOTAT	Museum of Transport and Technology (Auckland)
MOT	Ministry of Transport
MVA	Motor Vehicle Accident
NZFS	New Zealand Fire Service
PTO	Power Take-Off
RNZAF	Royal New Zealand Air Force
TTL	Turn Table Ladder
UFBA	United Fire Brigades Association
VFB	Volunteer Fire Brigade

TERMINOLOGY

Appliance	The general term for a fire engine. It covers both manual and mechanical machines. Sometimes referred to also as fire trucks or pumps.
First Aid Reel	The smaller hose reel fitted to most appliances and unwound to the fire. Normally used for small fires to limit the amount of water damage and to save time as it is permanently connected to the water supply.
Standpipe	A tubular fitting attached to the hydrant at one end and the hose at the other. It is often 'T' shaped and allows the water to be extracted from the mains. Modern types have controls on them.
Hydrants	Valves fitted to the water mains to allow the connection of the standpipe. The locations are usually indicated by a painted metal sign or a marking on the road.
Pumps	The mechanism which imparts the energy to the water to cause it to flow through the hose and be discharged on the fire. The term can also be used to define a certain type of appliance, as some specialist appliances are not fitted with 'pumps'.
Portable Pump	A pump that has its own separate power unit attached and can be removed and located closer to the water source or fire.
Wheeled Escape	A 16-metre extending ladder mounted on wheels. Its object was to rescue people trapped from multi-storied buildings. They were carried on appliances to the incident and man-handled into position.
Incident	The term used to describe each call-out by fire personnel.
Suction Hose	A hose specially designed to withstand negative pressures and used to gain water from non-pressurised sources, i.e. lakes, rivers, etc. This hose is coupled to the intake on the pump.
'Flaked' Hose	The term used when the normal hose is folded into a tray or locker in such a way that it can be removed quickly, sometimes while the appliance is moving.

Disclaimer

The obtaining of information for this book has been a major task. This is the first book of its type and covers the vast range of appliances used over almost 150 years. Very little information has been recorded and what has been, is published mainly by individual brigades who over the years have issued their own brigade histories. Overseas many major manufacturers have kept very accurate records, including photographs, but no such records exist in New Zealand.

Over the years a number of administrative changes have occurred: from Town Boards to Brigades; Brigades to New Zealand Fire Service and then, to complicate matters even more, the amalgamation of most local bodies into new city and district councils. All this has made research a nightmare.

We have accepted the information supplied to us in good faith; it is as accurate as we have been able to ascertain. As we intend to establish a data base for photographs and historical information, we would welcome any further information including that which will assist us to establish the accuracy of these details.

The address is: The Curator, Chariots For Fire, P.O. Box 11, Kaikohe, New Zealand

INDEX